Integrated Water Resource Management

Edson de Oliveira Vieira • Samuel Sandoval-Solis
Valmir de Albuquerque Pedrosa
J. Pablo Ortiz-Partida
Editors

Integrated Water Resource Management

Cases from Africa, Asia, Australia, Latin America and USA

Springer

Editors
Edson de Oliveira Vieira
Federal University of Minas Gerais
Montes Claros, Minas Gerais, Brazil

Valmir de Albuquerque Pedrosa
Federal University of Alagoas
Maceió, Alagoas, Brazil

Samuel Sandoval-Solis
Department of Land, Air and Water
Resources
University of California, Davis
Davis, CA, USA

J. Pablo Ortiz-Partida
Hydrologic Sciences Graduate Group
University of California, Davis
Davis, CA, USA

ISBN 978-3-030-16564-2 ISBN 978-3-030-16565-9 (eBook)
https://doi.org/10.1007/978-3-030-16565-9

This Springer imprint is published by the registered company Springer Nature Switzerland AG
The registered company address is: Gewerbestrasse 11, 6330 Cham, Switzerland

Preface

Integrated water resources management (IWRM) is a process toward a sustainable development of water resources. IWRM incorporates economic, societal, and environmental sectors across and along boundaries. IWRM potentializes the integration of sectors, users, and all related interconnections with water resources. Despite its increasing conceptual popularity, the complexity of water systems and their political, social, economic, and environmental features can catalogue the implementation and effectiveness, which are incipient but challenging. As noted in the different chapters of this book, water issues are diverse, and therefore solutions differ from one area to another highlighting the need to adapt the IWRM actions and tools to the personality of each of the river basin contexts.

Operational actions have been contributing to the process of IWRM. Even when implemented at a very specific level, it should be integrated within the management of the whole hydrographic basin. Some countries throughout the world have been implementing many projects with certain IWRM components, as could be seen in many chapters of this book, e.g., the participatory approach in Brazil, evaluation (social impact study) in Costa Rica, transboundary treaties between the USA and Mexico, or framework for adaptation to climate change in Mexico, among others. Such experiences provide other initiatives with a solid groundwork in several fields of water resources management. These projects are proceeding slowly but, according to the sociopolitical and financial capacities and the local context, are always keeping the fundamental IWRM principles in mind as framework and guideline.

A common characteristic is that renewal of management strategies and policy mechanisms always comes after a conflict or as means to adapt the set of instruments to tackle extreme climate events and prevent future sociopolitical and environmental damages. Through the chapters of this book, multiple sources of such conflicts or the lack of flexibility and adaptation on water systems management was exposed. For example, the disconnection of the surface and groundwater management is a major issue that needs to be addressed toward effective planning and implementation of an IWRM framework based on the specific local and broader context.

The experiences presented in this book show that the effective implementation of IWRM can take several decades. Success in some countries is still accompanied by continuous challenges. Some goals, such as reconciliation of human water needs with economic sustainability and ecosystem needs, require considerable changes in the current management process and in the water culture, which may lead to even greater time to achieve these goals. Given the short-term focus of decision-makers and policymakers in most areas, there is always the temptation to seek quick solutions and to abandon the IWRM process if immediate gains are insufficient. Thus, in order to achieve the ultimate goals of IWRM, besides being an approach broadly advocated by international organizations and regional and local communities of experts, IWRM is an ad hoc strategy to facilitate sustainable and adaptive water resources management across scales in the sociopolitical and environmental watershed conditions. Needless to say, the integrated water resources management throughout the world requires a good dose of political will in order to secure water and to foster environmental sustainability and socioeconomic prosperities.

This book will provide some case studies showing important experiences related with IWRM throughout the world bringing a case from Brazil, the USA, Mexico, Costa Rica, Chile, South Korea, Iran, and some countries with severe water shortage problems, such as in Africa. Chapter 1 presents theoretical concepts, basis, responsibilities, and challenges of IWRM, tools necessary for effective IWRM, and economic, social, and environmental conditions of a basin that are related with IWRM. Chapter 2 presents an analysis of policies and regulations for water management in Brazil showing the principles, instruments, and institutional arrangements (National Water Resources Council, catchment basin committees, water agencies, and other bodies and agencies of the federal, state, and municipal governments) that are molding water management in the country. Chapter 3 presents a necessity of IWRM to solve conflicts for water in São Francisco Basin in Brazil. Chapter 4 describes the drivers that guided the State of California toward adapting an integrated water resources management framework. Chapter 5 analyzes international regulations for water markets and water banking in Australia, Chile, and California. Chapter 6 reviews the implications of climate change for water resources systems in Mexico and evaluates how management strategies from California can serve as potential adaptation schemes toward an integrated water resources management framework in Mexico. Chapter 7 illustrates the potential to advance transboundary water resources management in a more comprehensive approach. The focus is given to the transboundary Paso del Norte (PdN) region which is considered as the most environmentally damaged, hydrologically developed, and prolific irrigated area in the Rio Grande/Rio Bravo (RGB) Basin. Chapter 8 intends to give a global overview of the situation of natural resources in Guanacaste, Costa Rica, where a ratio of the water resources is managed addressing the postmodern society in the region. Subsequently, the chapter unfolds with major conflicts that occurred in Guanacaste watershed over the last 20 years and the solutions implemented. In Chap. 9, the current status of water resources in Iran is reviewed through the study of two key critical cases in the country, Zayandehrud River Basin and Lake Urmia. In this chapter, challenges, management practices, and government policies are investigated. A new perspective is then drawn by the suggestion of implementing systems thinking and consideration of integrated water resources management opportunities. The Chap. 10 presents an overview of the current state of availability and the use of water resources, characteristics of rivers, large reservoirs, water quality management, and the future water resources management in South Korea. Chapter 11 presents the management and international water law instruments of transboundary groundwater in Africa. Transboundary aquifers represent an important source of water in Africa. Huge reserves of groundwater are located in some of the driest parts of this continent. Many of these watercourses and fossil aquifers are the subjects of state practices. This chapter shows few agreements including specific regulations to manage transboundary groundwater in Africa. Chapter 12 concludes with some considerations about the complexity of IWRM and its interrelationships between cultural, religious, and political aspects in different countries. This book will be of broad interest to professionals and students of hydrology and environmental science, politicians, stakeholders, and decision-makers in water resources.

Montes Claros, Minas Gerais, Brazil — Edson de Oliveira Vieira
Davis, CA, USA — Samuel Sandoval-Solis
Davis, CA, USA — J. Pablo Ortiz-Partida
Villahermosa, Tabasco, Mexico — Luzma Fabiola Nava

Contents

Contributors

Mohamad Azizipour Department of Civil Engineering, Faculty of Engineering, Shahid Chamran University of Ahvaz, Ahvaz, Iran

School of Civil Engineering, Iran University of Science and Technology, Tehran, Iran

Pável Bautista-Solís Mesoamerican Center of Sustainable Development of the Dry Tropics (CEMEDE), Universidad Nacional, Heredia, Costa Rica

Demétrius David da Silva Department of Agricultural Engineering, Federal University of Viçosa, Viçosa, Brazil

Valmir de Albuquerque Pedrosa Federal University of Alagoas, Maceió, Alagoas, Brazil

Edson de Oliveira Vieira Federal University of Minas Gerais, Montes Claros, Minas Gerais, Brazil

Erfan Goharian Department of Civil and Environmental Engineering, University of South Carolina, Columbia, SC, USA

Jesús Arellano-Gonzalez Agricultural and Resource Economics, University of California, Davis, Davis, CA, USA

Josué Medellín-Azuara School of Engineering, University of California, Merced, Merced, CA, USA

María E. Milanés Murcia Sacramento, CA, USA

Ricardo Morataya-Montenegro Universidad Nacional, Heredia, Costa Rica

Luzma Fabiola Nava Center for Global Change and Sustainability C.A. (CCGS), Villahermosa, Tabasco, Mexico

International Institute for Applied Systems Analysis (IIASA), Laxenburg, Austria

J. Pablo Ortiz-Partida Hydrologic Sciences Graduate Group, University of California, Davis, Davis, CA, USA

Silvio Bueno Pereira (Deceased) Department of Agricultural Engineering, Federal University of Viçosa, Viçosa, Brazil

Samuel Sandoval-Solis Department of Land, Air and Water Resources, University of California, Davis, Davis, CA, USA

J. Edward Taylor Agricultural and Resource Economics, Social Sciences and Humanities, University of California, Davis, Davis, CA, USA

Jaeeung Yi Department of Civil Engineering, Ajou University, Suwon-Si, South Korea

Sooyeon Yi Department of Landscape Architecture and Environmental Planning, University of California Berkeley, Berkeley, CA, USA

Abbreviations

AFB	African Development Bank
AGB Peixe Vivo	Peixe Vivo River Basin Management Support Executive Association
ALD	Alavijeh-Dehagh (sub-basin downstream of the Zayandehrud Dam)
ANA	Agência Nacional de Águas (National Water Agency)
ASADA	Asociaciones Administradoras de Acueductos Rurales
AyA	Instituto Costarricense de Acueductos y Alcantarillados
BCM/year	Billion cubic meters per year
BM	Boein-Miandasht (sub-basin upstream of the Zayandehrud Dam)
BOD	Biochemical oxygen demand
BS	Ben-Saman (sub-basin downstream of the Zayandehrud Dam)
CBHSF	São Francisco River Basin Committee
CCA	Water Advisory Council
CDS	Comisión sobre el Desarrollo Sostenible
CEBDS	Conselho Empresarial Brasileiro para o Desenvolvimento Sustentável
CEMIG	Energy of Minas Gerais Company
CEPAL	Comisión Económica para América Latina y el Caribe
CERH	Conselho Estadual de Recursos Hídricos (State Councils of Water Resources)
CHD	Chadegan (sub-basin upstream of the Zayandehrud Dam)
CHESF	Hydroelectric of São Francisco Company
CHGH	Chelgerd-Ghaleshahrokh (sub-basin upstream of the Zayandehrud Dam)
CHKH	Chel-Khaneh (sub-basin upstream of the Zayandehrud Dam)
CLD	Causal loop diagrams
CNRH	Conselho Nacional de Recursos Hídricos (National Council of Water Resources)

CODEVASF	Development Company of the São Francisco and Parnaíba valleys
CONAGUA	National Water Commission of Mexico
CONAMA	Conselho Nacional do Meio Ambiente
CR	Colorado River
CVP	Central Valley Project
DAD	Damaneh-Daran (sub-basin upstream of the Zayandehrud Dam)
EB	Elephant Butte Dam
ECOLEX	Environmental law database online
FAO	Food and Agriculture Organization of the United Nations
FIRO	Forecast Informed Reservoir Operations
GEAS	Global Environment Alert Service
GIS	Geographic information systems
GPD	Gross domestic product
HAR	Hydrological Administrative Regions
IAEA	International Atomic Energy Agency
IBC	International Boundary Commission
IBWC	International Boundary and Water Commission
IGRAC	International Groundwater Resources Assessment Centre
ILC	International Law Commission of the United Nations
INEC	Instituto Nacional De Estadística y Censos
IPCC	Intergovernmental Panel on Climate Change
IWRM	Integrated water resources management
KS	Kuhpaye-Sagzi (sub-basin downstream of the Zayandehrud Dam)
kV	Karvan (sub-basin downstream of the Zayandehrud Dam)
LAN	National Water Law
LGCC	General Law on Climate Change
LGEEPA	General Law of Ecological Balance and Environmental Protection
LJ	Lenjanat (sub-basin downstream of the Zayandehrud Dam)
MCM	Millions of cubic meters
MEIM	Meimeh (sub-basin downstream of the Zayandehrud Dam)
MUKH	Murche-Khort (sub-basin downstream of the Zayandehrud Dam)
NGO	Non-governmental organization
NJ	Najafabad (sub-basin downstream of the Zayandehrud Dam)
NM	New Mexico
NMHA	North-Mahyar (sub-basin downstream of the Zayandehrud Dam)
OECD	Organisation for Economic Co-operation and Development
ONS	System National Operator
OSS	Sahara and Sahel Observatory
PCH	Small hydroelectric power stations
PdN	Paso del Norte

PDNWC	Paso del Norte Watershed Council
PdNWTF	Paso del Norte Water Task Force
PE	Petrolina, Brazil
PISF	São Francisco River Integration Project
PND	National Development Plan
PNI	National Infrastructure Program
PNRH	Política Nacional de Recursos Hídricos (National Water Resources Policy)
PVWMA	Pajaro Valley Water Management Agency
Q7,10	Minimum flow of 7 consecutive days and return period of 10 years
Q90	Flow rate associated with flow permanence of 90%
Q90reg	Regularized flow rate associated with flow permanence of 90%
Q95	Flow rate associated with flow permanence of 95%
Q95reg	Regularized flow rate associated flow permanence of 95%
QLT	Long-term average streamflow
Qmo	Maximum water flow granted
Qmr	Minimum flows of reference
Qr	Minimum residual flows
RGB	Rio Grande/Rio Bravo
SADC	Southern African Development Community
SD	System dynamics
SINGREH	Sistema Nacional de Gerenciamento de Recursos Hídricos (National System for Water Resources Management)
SRH	Secretaria de Recursos Hídricos (Secretariat of Water Resources)
SRHU	Secretaria de Recursos Hídricos do Ministério do Meio Ambiente (Secretariat of Water Resources of the Ministry of the Environment)
SWP	State Water Project
TP	Total phosphorus
TR	Tijuana River
Tx	Texas
ULRP	Urmia Lake Restoration Program
UN	United Nations
UNECE	United Nations Economic Commission for Europe
UNECOSOC	United Nations Economic and Social Council
UNEP	United Nations Environment Programme
US	United States
USGS	United States Geological Survey
YCH	Yan-Cheshmeh (sub-basin upstream of the Zayandehrud Dam)

Chapter 1
Integrated Water Resources Management: Theoretical Concepts, Basis, Responsibilities, and Challenges of IWRM

Edson de Oliveira Vieira

Abstract This chapter describes the underlying theoretical concepts, the basics, and the responsibilities of IWRM. What principles guide the management and development of global efforts for the implementation of IWRM? This chapter also presents some tools needed for effective IWRM and how the economic, social, and environmental conditions of a basin are related to IWRM. What are the main governance and public roles in IWRM? This chapter identifies some of the key challenges of implementing IWRM.

Keywords Water management tools · Water management policies · Water governance

1.1 Introduction

Water is essential to life. It is present everywhere, but life cannot exist without its liquid form. Water plays a vital role in almost all human activities, including industry, agriculture, energy production, transportation, sanitation, navigability, and recreation, among others. Our planet has approximately 13×10^{13} million liters of water; however, 97% of this water is seawater, making it unfit for most human activities. Of the remaining 3%, only 0.4% is accessible and usable by human beings (Brutsaert 2005).

Water is constantly in motion in our planet, passing from one state to another, and from one location to another, which makes its rational planning and management a very complex and difficult task under the best of circumstances (Biswas 2004). Water may be everywhere, but its use has always been constrained in terms of availability, quantity, and quality. Population increase in cities with accelerating economic activities has been increasing water demand, energy production, and food, creating further pressures on the water resources (Setegn and Donoso 2015).

E. O. Vieira (✉)
Federal University of Minas Gerais, Montes Claros, Minas Gerais, Brazil
e-mail: eovieira@ica.ufmg.br

E. O. Vieira et al. (eds.), *Integrated Water Resource Management*,
https://doi.org/10.1007/978-3-030-16565-9_1

These factors of pressure are a source of conflicts that can vary significantly from one region to another or among season or even within a country. In addition, other drivers that put pressure on water availability include globalization of world economy, climate change, land use, demography, and urbanization. All these drivers have caused demand for water to increase drastically over the past century. The world population tripled during the twentieth century, while water withdrawals increased by a factor of seven (GWP 2000a). It is estimated that the world's population will increase by about three billion people by 2050. Much of this growth will take place in developing or middle-income countries like Brazil, South Africa, China, and India, with all challenges that carry in terms of investment needs for water supply and water treatment. Other developing countries already suffer water scarcity problems and lack the infrastructure and institutions needed to provide water services and manage water conflicts. Conflicts also exist among various water use sectors and societies, urban and rural water users, hydropower demand, environment, and irrigated agriculture, between upstream and downstream areas, and even between the same water sectors such as agriculture. Current and past approaches of water resources management have been proving inadequate to solve water conflicts and/or even for the global water challenges. These approaches are mostly sectoral management, where each sector (domestic use, agriculture, industry, sanitation, environmental protection, etc.) has been managed separately, with limited or inexistent coordination among sectors. These approaches lead to the fragmented and uncoordinated development of water resources. Thereby, integrated water resources management (IWRM) has appeared as a way of addressing local and global water problems to obtain a sustainable water management.

1.2 Mistaken Approaches to Water Resources Management

Water resources management (WRM) has been subjected over many years to an inadequate approach, considering the major challenges already foreseen for all sectors (domestic use, agriculture, industry, environmental protection, etc.). In most cases, these sectors presented management without a relation to each other, thus completely independent. This approach has resulted in a fragmented and uncoordinated development of water resources and has generated conflicts in many parts of the world. Water must be thought across the various sectoral boundaries, recognizing the interdependencies over the use of water in these sectors. As water becomes scarcer, it becomes increasingly inefficient to manage water without recognizing sectoral interdependencies, and even considering priority uses, conflicts can become difficult to solve (Xie 2006). Central governments, over the years, have adopted top-down approaches, centrally without prior consultation with water users or society. Such approaches dominate the processes of water resources management in many countries throughout the world and had questionable effectiveness. Central governments emphasized increasing supply relative to demand management, leading to an inefficient development project.

The low efficiency and quality of water supply services result in a vicious circle where dissatisfied users refuse to pay water charges, limiting the ability of these service providers to maintain infrastructure effectively causing a decline in quality of service. Poor service quality in turn exacerbates poor productivity of water and leads to the depletion of aquifers and pollution of water bodies. Artificially low water prices fail to encourage conservation and efficiency and allow wasteful practices and inefficient operations to continue.

When water resources management started to be considered, supply management has been predominant. Disregarding demand management as a priority, it has led to supply management to cause negative externalities, increasing the opportunity cost of water to unsustainable levels.

The growth of activities that require a lot of water, allied with the increase of the concentration of populations in cities, put pressure on already scarce water resources. As a result, new water sources need to be obtained, and larger reservoirs need to be built, resulting in greater ecological and social consequences. The problem of water scarcity is often the result of a crisis of management or governance rather than to considerate only an imbalance between input and output of water into the physical system. Failure to meet social and environmental demands, the ineffectively regulated pollutant load, the inefficiency of water service providers, and the fall of allocation of scarce water resources are examples of this crisis of governance of water resources. Only a change in the way water resources are managed can prevent an even worse water crisis.

The shortcomings mentioned above with traditional WRM approaches triggered the development of an IWRM framework that has emerged as a means of addressing global water problems and working toward a sustainable future for water management (Xie 2006).

1.3 Definition of Integrated Water Resources Management

Certainly, to improve the water resources management process, there is recognition of the need to implement a more holistic approach to water management than has been practiced in the past. However, there is no consensus on the definition of IWRM and what implies the implementation of an IWRM approach (Bateman and Rancier 2012).

Some few members of the water profession started to realize during the 1980s that the water resources management throughout the world is not as good as they appeared. This feeling intensified during the 1990s when many in the profession began to appreciate that the water problems have become multidimensional, multisectoral, and multiregional and filled with multi-interests, multi-agendas, and multicauses, which can be resolved only through a proper multiinstitutional and multi-stakeholder coordination (Biswas 2004).

An international organization dedicated to promoting sustainable management of water resources, the Global Water Partnership, defined the term integrated water resources management (IWRM) as "a process which promotes the coordinated development and management of water, land and related resources, in order to maximize the resultant economic and social welfare in an equitable manner without compromising the sustainability of vital ecosystems" (GWP 2000b).

In a survey about IWRM of more than 600 professionals in the United States, it was described as "a process that strives to balance regional economic growth while achieving wise environmental stewardship by encouraging the participation of seemingly disparate interests" (Bourget 2006).

The position statement of the American Water Resources Association (AWRA) (Bateman and Rancier 2012) identifies IWRM as "The coordinated planning, development, protection, and management of water, land, and related resources in a manner that fosters sustainable economic activity, improves or sustains environmental quality, ensures public health and safety, and provides for the sustainability of communities and ecosystems."

Based on results from research during a series of regional conferences, the US Army Corps of Engineers (USACE) defined IWRM as "IWRM aims to develop and manage water, land, and related resources, while considering multiple viewpoints of how water should be managed (i.e. planned, designed and constructed, managed, evaluated, and regulated). It is a goal-directed process for controlling the development and use of river, lake, ocean, wetland, and other water assets in ways that integrate and balance stakeholder interests, objectives, and desired outcomes across levels of governance and water sectors for the sustainable use of the earth's resources" (USACE 2010).

The IWRM shouldn't be seen as an end but as a means to achieve three strategic targets:

- Efficiency in the use of water and other related natural resources.
- Equity in the allocation of water resources among different socioeconomic groups.
- Social, economic, and environmental sustainability to protect water resources and associated ecosystems.

1.4 IWRM at the Policy Level

1.4.1 Water: Scarcity or Mismanagement?

Most of water managers throughout the world know that water scarcity results from a crisis of governance. The lack of water policies or even inadequate water management sometimes results in tragic effects on poor populations around the world. In seeking to implement IWRM, it is necessary to recognize some key criteria that consider social, economic, and natural conditions.

1.4.1.1 The Watershed as Environmental Spatial Units

Watersheds are spatial units of varied dimensions where the water resources are organized as a function of the relations between the geomorphological structure and the climatic conditions. IWRM considers the watersheds as the basic water management unit to be a conjunction of environmental factors. Watersheds are understood as basic cells of environmental analysis, where the systemic and integrated view of the environment is implicit. The environmental components such as rocks, relief, soils, water, vegetation, and climate could no longer be understood separately, but it would be fundamental to recognize their interfaces and interconnections to understand the environmental dynamics and propose a sustainable planning and management of the ecosystems. The international agreements and processes relative to climate change; desertification; biodiversity; arid, semiarid, and humid zones; etc. could be the basis for the introduction of new environmental action policies; but their efficient implementation requires that they be viewed in the context of the sustainable management and regeneration of all-natural resources. It follows from the systemic conception that water should not be managed without considering its close interrelations with the other components of the environment, e.g., climate, soils, geology, vegetation, relief, and anthropic action, that changes the working conditions of natural systems, producing changes that can directly affect the quality and quantity of water available in a basin.

1.4.1.2 Social and Institutional Aspects: Participation and Decentralization

To ensure the sustainability of water resources, IWRM underlines the importance of involving all stakeholders within watershed: the governmental authorities, public and private institutions, public and private sectors, and civil society, with a special focus on women and marginalized groups. Decentralized participation is understood as an instrument to enable and legitimize public policies that intervene in the water management system. In principle, the participation of different segments of civil society, representing interests of different water users and citizens, from the elaboration to the implementation of plans and projects, would tend to generate more equitable, effective, and legitimate decisions, plans, actions, and projects. The structure of this framework should correspond to local sociocultural, ecological, and economic conditions. Local participation should be backed by close cooperation at higher institutional levels: between the agencies, departments, and ministries that administer water, agriculture, the environment, industries, etc. In this way participation and decentralization can maintain the priorities of the majority over some isolated interests, even economically preponderant.

1.4.1.3 The Economic Aspect

How can rentability be increased without penalizing the poor? International organizations such as the World Bank and the International Monetary Fund (IMF) propose to privatize the water sector, arguing that this would eliminate monopolies and abusive prices. The issue is controversial however if on the one hand the privatization could give rise to new forms of power and dependencies linked to a service that the population cannot live without it and on the other hand the lack of control of this trade for social control may jeopardize an essential human right, that is, the access to drinking water by the poorest population. Some ideas have been formulated: free provision of the quantity of water for living (30–50 liters per person per day according to the World Health Organization) and adjusting water rates to income, a price that would be inversely proportional to the distance people must cover to meet their water needs. This subject is of great complexity, and much must be discussed before deciding on the marketing of water. Local and regional aspects should be considered in this discussion.

1.5 The Dublin Principles

In 1992, the International Conference on Water and Environment (ICWE) held in Dublin, Ireland, more than 500 participants representing 100 countries and 80 international and nongovernmental organizations, according to the level of policy of WRM, recommended four principles to guide global effort management and development (these four principles were adapted from Cap-Net (2010), GWP (2017), and Xie (2006)):

Principle 1: "Ecological" – *Fresh water is a finite and vulnerable resource, essential to sustain life, development, and the environment.*

- Water sustains life in all its forms, being a necessary resource for different purposes, functions, and services. It is for this reason that holistic and integrated water management must consider the demands and threats on resources (in this case not only water but everything related to it). Integrated management involves not only the management of natural resources but also involves coordination between different human activities that need water for different uses, linking social and economic development with protection of natural systems. In addition, it is necessary to determine the different uses of the soil and identify those that produce waste that can contaminate the water. It should be stressed that the creation of a political system sensitive to water issues requires the coordination of policies and institutions at all levels (from national ministries to local authorities or the community). There is also a need for mechanisms to ensure that decision-makers consider the costs associated with water use when making domestic production and consumption decisions. The development of an institutional framework covering all aspects mentioned above and capable of integrating human, economic, social, and political resources presents a considerable challenge. This principle recognizes the watershed or river basin as the most appropriate unit for water governance and calls for coordination across the range of human activities that use and affect water in a given river basin. IWRM approaches incorporate this principle into its emphasis on integration between all concerned water sectors.

Principle 2: "Institutional" – The development and management of water resources must be participatory, involving at all levels those who plan, use, and decide.

- We are all stakeholders when it comes to water use. Effective participation in water resources management only happens when everyone is part of the decision-making process. It is to raise awareness of water issues among policy-makers and the public. Management decisions should be taken at the lowest appropriate level. This can happen at the local level as communities come together to make decisions about water supply, planning, management, and water use. Participation may be at the regional level with the democratic representation of people elected by stakeholder groups. In any case, the type of participation in decision-making in water management will depend on the magnitude of the project or program, the technical knowledge, the necessary investments, and the economic and political system concerned. This principle advocates increased accountability of management institutions and full consultation and involvement of users in the planning and implementation of water projects. The capacity of certain disadvantaged groups may need to be enhanced through training and targeted pro-poor development policies for full participation.

Principle 3: "Gender" – Women play a central role in water supply, management, and safeguarding.

- This principle emphasizes the important synergy that exists between gender equity and sustainable water management. It is well known that in many countries women play a key role in collecting and safeguarding water, for various purposes, mainly for domestic and agricultural purposes. However, in many societies, women are excluded from water management decisions. To consider gender as a crosscutting objective in the development of water policy requires recognition of the role of women, their ideas, and their interests and needs, in the same way that men's views are recognized. Development policies, particularly water management, should support equal rights and responsibilities between women and men. It is for this reason that gender must be considered when developing or updating the legal framework to ensure that policies, programs, and projects address different experiences and situations between women and men. Equitable participation in social and political aspects means that women have the same right to express their needs and interests as well as their vision of society, shaping the decisions that affect their lives. One way to enhance the capacity for equitable participation is through community organizations and related institutions. IWRM includes an emphasis on empowering women in its focus on participatory management and capacity building.

Principle 4: "Economic" – Water has an economic value in all its competing uses and should be recognized as an economic and social good.

- Water has a value as an economic good as well as a social good. Many failures of water resources management in the past may be related to the nonrecognition of water with economic value. Having access to safe drinking water and sanitation at an affordable price is a basic right of all human beings, and this should be recognized by everyone. The nonobservance or recognition that water has economic value has led to inappropriate uses of this resource and harmful to the environment with very high water waste by stakeholders. Water management as an economic good is an important means of achieving efficient and equitable use, as well as encouraging the conservation and protection of water resources. Value and price are two distinct concepts. The value of water for alternative uses is important for the rational allocation of water as a scarce resource, either by regulatory or economic means. On the other hand, the price of water is related to the application of an economic instrument to achieve multiple objectives: supporting disadvantaged groups, influencing water conservation, increasing and stimulating the efficiency of water use and demand management, and securing costs and consumers willing to pay additional investments in water services. Managing water as an economic good is also a key to achieving financial sustainability of water service provision, by making sure that water is priced at levels that ensure full cost recovery. IWRM emphasizes on economic and financial sustainability.

1.6 IWRM Tools

There is no specific model to be adopted for the implementation of the IWRM due to the high degree of complexity and specificity existing in the water management of each country. Thus, the Global Water Partnership has created an IWRM ToolBox designed to support the development and application of IWRM in many situations. These tools will help in the adequacy of the implementation of the IWRM according to the specific situation analyzed and according to their needs. The tools fall into three overarching pillars: (a) enabling environment, (b) institutional roles, and (c) management instruments. Each pillar has several subcategories, which, in turn, consist of several tools, with 62 tools in total, but below is shown the first two levels of each pillar (GWP 2017):

- (a) Enabling environment: This pillar is subdivided into three subcategories that must be established to achieve a sustainable balance between the social, economic, and environmental needs of water:
  - (a1) Policies that define national and regional objectives incorporating the concepts of integration, decentralization, participation, and sustainability of the IWRM, to establish water use, protection, and conservation goals.

  - (a2) Legislative framework to translate water policy into law covering water ownership, licenses, and water use rights and the legal status of water user groups.
  - (a3) Funding/incentives: Financing and incentive structures are needed to fund capital-intensive water projects, support water service delivery, and provide other public goods such as flood control and preparedness for period of water scarcity or severe drought. This source of funding can be resources from the public sector, private finance, and joint public-private partnership. The enabling environment facilitates all stakeholders to play their respective roles in the sustainable and management of water resources.

- (b) Institutional arrangements. This pillar consists of four subcategories:
  - (b1) Regulation and compliance: Constituted by the set of agencies and governmental and private institutions for the execution of the policy, through an organizational structure to be adopted aiming at integrated, decentralized, and participatory management. These organizations need to have well-defined rights and responsibilities and allow integration among them.
  - (b2) Water supply and sanitation services: Institutions of water supply and sanitation services can be public, private, or cooperatively owned and managed entities but can also result from collaborations between these sectors, such as public sector water utilities, private sector water service providers, and community-based water supply and management organizations.
  - (b3) Coordination and facilitation: The main role of the coordination and facilitation bodies is to articulate and harmonize the actions and visions of the many entities involved in water management by putting the actors involved around the same table and guiding them toward a collective goal and vision.
  - (b4) Building institutional capacity: All actors that are an integral part of the water resources management process must be capacitated and trained in the skills and instruments of effective water management and in accordance with IWRM principles. Human resources development through training, education, and provision of information is a key dimension of capacity building.

- (c) Management instruments. Once the proper enabling environment and institutions were implemented and have been working, these instruments address specific management problems adopting detailed methods that enable decision-makers to make rational and informed choices between alternative actions when it comes to water management. These choices should be based on agreed policies, available resources, environmental impacts, and the social and economic consequences. Quantitative and qualitative methods are being combined with a knowledge of economics, hydrology, hydraulics, environmental sciences, sociology, and other disciplines pertinent to the problem in question, for defining and evaluating alternative water management plans and implementation schemes.

  - (c1) Understanding water endowments: Management of water resources requires understanding resources and needs, demands, and supplies, identifying and listing priority areas, monitoring and evaluating systems, and involving data collection and analysis to inform decision-making with a holistic view of water resources and its interaction with societal use in a country or region.

- (c2) Assessment instruments: Help to understand the connections between water resources and their multiple users as well as to calculate the impacts of uncertain events or policy measures on the resource and its users. The aspects considered are risk and vulnerability, social structures and effects, ecosystems, environment, and economics.
- (c3) Modeling and decision-making: Sustainable management of water resource requires a good understanding of the distribution and quantities of that resource. Thus, information is very important, but it can be hard to obtain and to manage. A good management of water resources requires a huge and reliable amount of spatially and temporally varying data from many different sectors: the quality and quantity of water resources; the geography of the area; land use, soil, and local geology; and the human communities. Analytical tools are needed to interpret the data in a way that makes it usable for decision-makers. Models as geographic information systems (GIS) and decision support systems (DSS) do exactly that.
- (c4) Planning for IWRM: IWRM plans are one of the key pillars of integrated water management, identify actions and a set of management instruments that are embedded in a wider framework of policies, legislation, financing structures, and capable institutions with clearly defined roles and should involve social participation in its building process.
- (c5) Efficiency in water management: Water demand management and water supply management constitute an important instrument of IWRM. Efficient use of water, improving supply and demand efficiency, increasing of water reuse as well subsidies, and the regulation to encourage technology improvements are important strategies in IWRM implementation/practice.
- (c6) Communication means exchanging information, and this instrument is fundamental to the success in IWRM. Communication allows different sectors that use water resources to share information and collaborate on management issues. Communication allows involvement of stakeholders in the decision-making and implementation process. All parties involved should maintain effective communication in relation to water management.
- (c7) Economic instruments: Are one way to promote changing the behavior of water users toward more sustainable practices. Economic incentives involve the use of prices and other market-based measures to improve the way water is managed and used. They provide incentives to rational water use, efficiently and in a manner consistent with the public interest. They have both positive and negative effects, rewarding users that recognize the true value of water and penalizing profligate and antisocial use.
- (c8) Social inclusion of the most deprived social groups, promotion public awareness for water issues, stakeholder participation in water planning and operating decisions, teaching more sustainable water use practices for children and youth in school, and externalization of water footprint or virtual

water are important tools for social change and a necessary step in achieving water security.

The IWRM tools shown above illustrate the multifaceted approach by IWRM and how hard it is to put the Dublin Principles into practice. The implementation of the IWRM at the national, regional, and local level requires application of some or many of these tools in a manner complementary and simultaneous application of several other tools. Before the implementation of tools, they should be carefully evaluated and selected those which fit each context. It is advisable to monitor and evaluate changes as tools are adopted to prevent unintended or undesirable consequences.

1.7 Challenges in the Practical Application

The main challenge of the IWRM is to meet the four principles proposed by the International Conference on Water and Environment (ICWE) held in Dublin in 1992. However, other challenges can be pointed to the implementation of the IWRM (Fulazzaky 2014; Garcia 2008; McDonnell 2008; Rahaman and Varis 2005). Regionally adapted integrated approaches are the following:

- Sustainable and optimal distribution and uses of water resources (surface and groundwater) without quantitative-qualitative overuse, considering ecological functions of water resources.
- Define integrated political actions adapted to regional conditions and cultures involving coordination with other areas at government levels.
- Plan and manage water resources for the distribution and multiple use of water, meeting its multiple objectives including economic, social, and environmental aspects.
- Increase water use efficiencies across sectors by dramatically reducing waste and increasing water availability.
- Establish qualitative and quantitative information systems involving GIS-linked databases to store and manage data from a river basin, helping at the operational level of this information.
- Incorporate irrigation management into water management in its three dimensions: scarcity, excess, and quality.
- Establish tools and legal framework to face climate change.
- Balancing productive development with the human right of access to water and preservation of the ecosystem.

1.8 Conclusions

Although the IWRM concept has become more popular in recent years, its implementation has been incipient. The nonuse of sustainable and integrated water management has made it difficult to cope with problems that are becoming increasingly complex. Operational action associated with management tools can contribute to the integrated management process. Even when implemented at a very specific level, it must be integrated into the management of the entire river basin. But successful actions in one country may not be in others. Traditions, customs, and other local, regional, or national cultural aspects should be considered in the implementation of IWRM actions and tools. Policies and plans from other water sectors as energy, agriculture, and forestry should integrate to IWRM. There are several projects with certain components of IWRM, already implemented around the world, such as participatory approach, evaluation, or financial aspects, and some will be presented in this book.

References

Bateman B, Rancier R (2012) Case studies in integrated water resources management: from local stewardship to national vision American Water Resources Association Policy Committee 4

Biswas AK (2004) Integrated water resources management: a reassessment: a water forum contribution. Water Int 29:248–256

Bourget PG (2006) Integrated water resources management curriculum in the United States: results of a recent survey. J Contemp Water Res Educ 135:107–114

Brutsaert W (2005) Hydrology: an introduction. Cambridge University Press, New York

Cap-Net (2010) Integrated Water Resources Management (IWRM) Tutorial. http://www.cap-net.org/wp-content/uploads/tutorials/iwrm_tutorial/mainmenu.htm. 2017

Fulazzaky MA (2014) Challenges of integrated water resources management in Indonesia. Water 6:2000–2020

Garcia LE (2008) Integrated water resources management: a 'small' step for conceptualists, a giant step for practitioners. Int J Water Res Develop 24:23–36

GWP (2000a) Background paper no. 4. Integrated water resources management, vol 4. Global Water Partnership, Technical Advisory Committee (TAC), Stockholm, Sweden

GWP (2000b) Towards water security: a framework for action. Global Water Partnership, Technical Advisory Committee (TAC), Stockholm, Sweden

GWP (2017) IWRM ToolBox teaching manual. Global Water Partnership, Stockholm, Sweden

McDonnell RA (2008) Challenges for integrated water resources management: how do we provide the knowledge to support truly integrated thinking? Int J Water Res Develop 24:131–143

Rahaman MM, Varis O (2005) Integrated water resources management: evolution, prospects and future challenges. Sustain Sci Pract Policy 1:15–21

Setegn SG, Donoso MC (2015) Sustainability of integrated water resources management: water governance, climate and ecohydrology. Springer International Publishing, Switzerland

USACE (2010) National report: responding to National Water Resources Challenges, building strong collaborative relationships for a sustainable water resources future. United States Army Corps of Engineers, Washington, DC

Xie M (2006) Integrated water resources management (IWRM)–introduction to principles and practices. In: Africa Regional Workshop on IWRM, Nairobi, Oct, 2006

Chapter 2
Integrated Water Resources Management in Brazil

Demétrius David da Silva, Silvio Bueno Pereira, and Edson de Oliveira Vieira

Abstract The process of management of water resources in Brazil is incipient and was established through the National Water Resources Policy (Política Nacional de Recursos Hídricos (PNRH)). The PNRH presents the foundation and principles of IWRM established in Dublin in 1992 and has good management instruments, but it is not fully implemented in Brazil. The PNRH gives priority to quantitative aspects and almost does not refer to groundwater. Cultural and regional characteristics have not been considered in the policy even though there is high diversity. Such aspects should be implemented in the basin water plans by the responsible basin committees. There is still much to be done to establish IWRM in Brazil.

Keywords Brazilian Policy of Water Resources · Brazilian Water Plan · Water management instruments

2.1 Introduction

Water plays different roles among the environmental resources. Sometimes water is seen as a product for direct consumption, as feedstock, or as an ecosystem. The main problem is that water has no substitute for many of its applications, such as human, plant, and animal consumption. Water has four main functions: (1) biological, as water for basic human and animal needs; (2) ecosystemic, to support aquatic and riparian species; (3) technical, when water is used as raw material in the generation of electric energy, in industry and agriculture, or in nonbasic residential uses; and (4) symbolic, which is associated with social and cultural values (Kemper et al. 2007).

Silvio Bueno Pereira died before publication of this work was completed.

D. David da Silva (✉) · S. B. Pereira (Deceased)
Department of Agricultural Engineering, Federal University of Viçosa, Viçosa, Brazil
e-mail: demetrius@ufv.br

E. O. Vieira
Federal University of Minas Gerais, Montes Claros, Minas Gerais, Brazil

E. O. Vieira et al. (eds.), *Integrated Water Resource Management*,
https://doi.org/10.1007/978-3-030-16565-9_2

In the current stage of human infrastructure development, there has been an intense deterioration of water quality in much of our planet. Considering the limited water resources, the situation is very worrying. Although water is a renewable resource through the physical processes of the hydrologic cycle, pollution compromises the fraction of the water that can be used as well as the discrepancy in spatial and temporal distribution of water reserves (ANA 2015b). This fact is visible in several regions in Brazil, where usually perennial streams are now intermittent stream, aggravating the problem of water scarcity. In addition, there is a greater tendency for extreme events to occur, with high values of maximum flows in the rainy season and extreme droughts in dry periods. Such dry periods are a particular concern for irrigated agriculture because it is during the dry season of the year that the greatest water demands occur.

In the current trend of global climate change, special attention has been paid to the most recent extreme events to assess whether these events, particularly droughts and floods, are associated with interannual variability or if they are because of climate change. For example, in the Amazon region, five of the ten largest floods observed since 1902 in the city of Manaus, along the Rio Negro, occurred after 2009. In a similar manner, four of the ten driest years occurred between 1997 and 2014.

In view of this new reality in Brazil, interest in the adequate use of water is increasing, and, consequently, the studies and actions related to IWRM are amplified. The evidence is the emergence of specific laws for this purpose, which have come to catalogue water as a scarce and finite resource as well as assigned it an economic value.

2.1.1 Surface Water Availability in Brazil

Although Brazil presents a privileged situation in relation to the world's water availability, with 13.8% of the fresh water of the planet (ANA 2009), and with a superior water availability per capita than most countries in the world, according to the UN (2017), the distribution of water is not uniform throughout the nation. The Amazon basin, which is inhabited by only 5% of the Brazilian's population and with a relative reduced consumptive demands, accounts for about 80% of water availability (ANA 2015a). Consequently, only 20% of the country's water resources are available in other regions, with more than 90% of the Brazilian population, where the greatest demands for water use occur.

To evaluate the surface water reserves, it is very important to understand the spatial and temporal distribution of precipitation. Precipitation varies between 1003 and 2205 mm in the São Francisco and Amazônica regions, respectively (Fig. 2.1, Table 2.1) (ANA 2013).

It should be noted, however, that the values presented in Table 2.1 are average in each hydrographic region and that the effective range of annual mean precipitation

Fig. 2.1 Hydrographic regions of Brazil. (Source: Adapted by authors from ANA (2013))

in Brazil is significantly higher, with values in the 500 mm range in the semiarid region of the northeast or 3000 mm in the Amazon region (ANA 2013).

In addition to the large spatial variability of precipitation, another relevant aspect is the temporal distribution of precipitations (seasonality). In Brazil it is common to associate sites with lower annual precipitation with practically all-rain events occurring in only 2 or 3 months of the year. This further aggravates the problem of water scarcity in these regions, as a significant part of the precipitation ends up as runoff, not recharging the aquifers. As a consequence, the groundwater table goes below the river channel and is not able to feed the watercourses in the dry season of the year. This is one of the reasons why most watercourses in the northeastern region of Brazil are intermittent in a natural regime, presenting only runoff during the wetter periods of the year.

To quantify the water availability of Brazil in its different hydrographic regions, ANA (2013) worked with long-term average streamflow data (Q_{LT}), characterized as the natural flow that would occur in a river basin without any human interference. Such human interferences are abstractions for uses and the minimum flow rate

Table 2.1 Average annual rainfall in the different hydrographic regions of Brazil from 1961 to 2007

Hydrographic regions	Total annual rainfall (mm)
Amazônica	2205
Tocantins-Araguaia	1774
Atlântico Nordeste Ocidental	1700
Parnaíba	1064
Atlântico Nordeste Oriental	1052
São Francisco	1003
Atlântico Leste	1018
Atlântico Sudeste	1401
Atlântico Sul	1644
Paraná	1543
Uruguai	1623
Paraguai	1359
Annual average rainfall	1761

Source: Adapted from ANA (2013)

Table 2.2 Long-term average streamflow (Q_{LT}) and flow rate associated with flow permanence of 95% (Q95) in the different hydrographic regions of Brazil

Hydrographic regions	Q_{LT} ($m^3\ s^{-1}$)	Q_{95} ($m^3\ s^{-1}$)
Amazônica	132.145	73.748
Tocantins-Araguaia	13.799	5.447
Atlântico Nordeste Ocidental	2.608	320
Parnaíba	767	379
Atlântico Nordeste Oriental	774	91
São Francisco	2.846	1.886
Atlântico Leste	1.484	305
Atlântico Sudeste	3.167	1.145
Atlântico Sul	4.055	647
Paraná	11.831	5.956
Uruguai	4.103	565
Paraguai	2.359	782
Annual average Brazil	179.938	91.271

Source: Adapted from ANA (2013)

associated with the 95% flow (Q_{95}), obtained from the permanence or durations curve elaborated on the basis of daily flow data (Table 2.2).

Table 2.2 shows that the most critical regions in relation to water availability in the driest period of the year, characterized by Q_{95}, are Atlântico Nordeste Oriental, Atlântico Leste, Atlântico Nordeste Ocidental, and Parnaíba. Particular attention is paid to the Parnaíba region, since in this case Q_{LT}, which theoretically represents the maximum flow rate that can be regularized in a basin for multiple uses, is only twice Q_{95}, while in the other three regions, this ratio ranges from 4.9 to 8.5. In the case of the São Francisco region, despite the high value of Q_{95} compared to the other regions of Brazil, it is worth noting that Q_{LT} is only 51% higher than Q_{95}.

2.2 National Water Resources Policy and the Insertion of Integrated Water Resources Management

Since the beginning of the last century, Brazil has begun to worry about the management of water resources, culminating with the promulgation in 1934 of the Water Code (Decree 24.643). The main motivations of this code were the lack of adequate legislation for the time, which wasn't in agreement with the needs and interests of the nation, and the need to endow the country with laws that would allow the public power to manage the use of water.

The Water Code presented a surprising vision of the future and constituted an extremely advanced legislation for the time. The code established a legal order for the use of the waters and clear norms for the use of watercourses and promotes the generation of water resources management instruments that are still in use.

In 1988, the Federal Constitution established significant changes in relation to water resources in its Article 21, item XIX, defining the National System for Water Resources Management (Sistema Nacional de Gerenciamento de Recursos Hídricos (SINGREH)) and defining the criteria to grant water rights. It also introduced new concepts, such as the federal or state dominance of waters and attributions related to water resources management in Brazil. The Federal Constitution stablished the goods of the Brazilian Government: lakes, rivers, and any other watercourses on lands in their domain or that to across more than one state or serve as limits with other countries or extend to foreign territory or from it, as well as marginal lands and river beaches. Also it establishes that the surface or groundwater and fluent, emerging, and deposited waters, in this case, under the terms of the law, are goods of the Brazilian Government.

Federal Law No. 9,433/1997, also known as "Water Law," incorporated the principles IWRM established in Dublin in 1992. This led to the migration of a centralized management model to a decentralized model, with the expectation of joint participation in the decision-making process of the governmental and nongovernmental segments (water users and civil society organizations). It also supported decision-making focused on collegiate instances of water resources, such as water councils and river basin committees.

Law No. 9,433/1997 established the National Water Resources Policy (PNRH) and created the National System for Water Resources Management (SINGREH). The implementation of this law changed the rules of water use in Brazil, especially in rural areas. Previously, if a producer decided to irrigate a certain area on his property, it would be sufficient for him to install a pumping set to capture the surface or underground water that was needed to meet his demand, without requiring any kind of authorization. However, since 1997, this situation has changed significantly because it became mandatory for the water user to obtain a permit to use water. The permit is an instrument issued by federal or state authority that authorizes the use of a certain amount of water for a certain period of time. In addition, in some cases, the user will have to pay for water use, and this resource will primarily be reverted to the river basin itself in actions aimed at improving water availability in quantitative and qualitative aspects. However, this payment refers to water as an economic good (not water pricing) and as an instrument to achieve multiple objectives such as influencing water conservation and increasing and stimulating water use efficiency and demand management.

Law 9,433/1997 adopts modern bases for IWRM as set out in the Dublin Declaration (International Conference on Water and the Environment, ICWE 1992) and Agenda 21 (United Nations Conference on Environment and Development 1992), among which are the priority use of water resources for human consumption and animal watering; the multiple uses of water; the adoption of the hydrographic basin as a physical-territorial planning unit; decentralized and participatory management, with the participation of public authorities, users, and civil society; and the recognition of water as a public domain property and as a limited natural resource with economic value (Brasil 1997).

The objectives of the PNRH are to ensure water availability, with the required quality standards for different uses, to current and future generations; the rational and integrated use of water resources, including water transport, with a view to sustainable development; and the prevention and defense against critical hydrological events of natural origin or resulting from inappropriate use of natural resources (Brasil 1997).

Among the resources management instruments foreseen in the PNRH (Brasil 1997), the following stand out:

- Water resources plans
- The framework of bodies of water into classes according to the prevailing uses of water
- The grant of rights to use water resources
- Charging for the use of water resources
- The information system on water resources

The National System for Water Resources Management (Sistema de Informação de Gestão de Recursos Hídricos (SINGREH)) is based on the concepts of IWRM and Law No. 9,433/1997 and the adoption of the river basin as territorial management unit, decentralized management, and the participation of public power, users, and communities in the process of deliberation on this management, being constituted by:

- National Water Resources Council (Conselho Nacional de Recursos Hídricos (CNRH))
- Water resources councils of the states and the federal district.
- River basin committees
- National Water Agency (Agência Nacional de Águas (ANA))

- Governing bodies of the federal and state governments, whose competencies are related to the management of water resources
- Watershed water agencies

2.2.1 *Water Resources Management Domains*

The multiple dominance of water bodies in a single basin, provided in Law 9,433/1997, requires the harmonization of federal and state laws, as well as the norms and procedures of the different water resources management agencies to formulate the river basin as a management unit. But, unfortunately, in practice this has not happened in most cases, and the management of water resources has been done in a segmented way, and the river basin is not effectively adopted as the basis for the IWRM.

Aiming to elucidate this issue, we exemplify some of the developments resulting from this multiple dominance in the grant of water use. Depending on the legislation, the reference flow values to be used for grant purposes may vary, since each state has the autonomy to adopt specific criteria for the establishment of minimum reference flows.

The dominance of watercourses in Brazil ends up fragmenting the management, since the National Water Agency and the different management bodies of the states and the federal district have different minimum flows of reference (Q_{mr}) and proportional percentages (Fig. 2.2). In addition, different forms are used for grant application, with different requirements regarding the hydrological information and studies to be presented, when the river basin should be effectively the planning and management unit.

It should be emphasized that the water resources management agencies in Brazil have the autonomy to determine the percentage of Q_{mr} ($Q_{7,10}$, Q_{90}, or Q_{95}) over which water rights are granted. In the federal case, for example, the maximum water flow (Q_{mo}) granted by the National Water Agency corresponds to 70% of Q95 and, consequently, minimum residual flows (Qr), downstream, of 30% of Q_{95} after the grant of all consumptive uses. It is also worth noting that many times this value of the remaining flow is called ecological flow, but this constituted a serious conceptual error, because in establishing this percentage of flow that must remain in the watercourses in the most critical periods does not take into account the actual needs of the river ecosystem in terms of flows and, therefore, shouldn't be called ecological flow.

The ecological flow corresponds to the amount of water that must remain in the watercourse in order to maintain the activities of the aquatic and riparian organisms. In order to determine this, besides the studies of the hydrological conditions of the basin, the analysis of the response of the aquatic species to the changes of hydrological factors should be done. The understanding of the interrelation biota-flow is essential for determining the ideal flows to support river ecosystems while considering the various activities and purposes that are desired with IWRM.

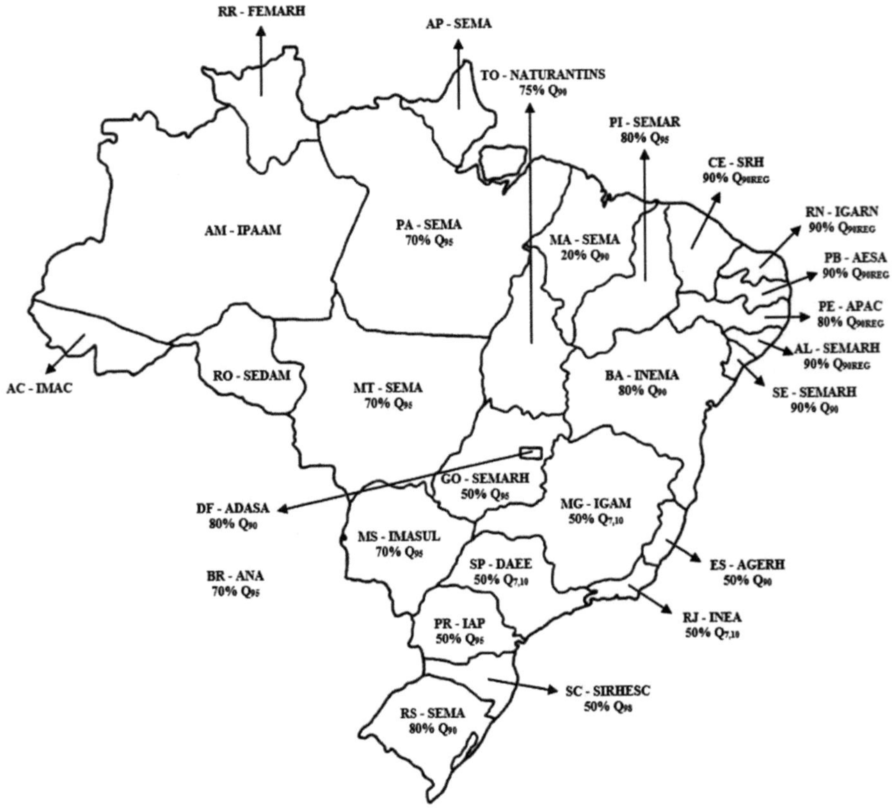

Fig. 2.2 Governing bodies of water resources in Brazil and respective percentages of minimum reference flows adopted for the purpose of grant water use. (Source: Adapted by authors from ANA (2013))

2.2.1.1 Water Resources Management Instruments

Water Resources Plans

In view of the legal grounds set forth in Law No. 9,433/1997, which defines the National Water Resources Policy, the water resources plans must present a minimum content that will inform and guide the implementation of this policy, considering the hydrographic basin as the unit of study and planning.

The water resources plan must have a technical content, sufficiently clear to allow its analysis by political leaders and financial agents, in order to make feasible the implementation of the programs and actions prioritized for the implementation of the plan.

Framework of Bodies of Water in Classes of Use

Law No. 9,433 related the framework of water bodies as one of the PNRH management instruments. The current guidelines and parameters for the classification of water bodies into classes of use were established by the National Council for the Environment (Conselho Nacional do Meio Ambiente (CONAMA)) Resolution No. 357 of March 17, 2005. The classification of watercourses into classes aims to ensure water quality compatible with the most demanding uses to which they are intended and to reduce the costs of combating water pollution by means of permanent preventive actions. This instrument makes it possible to relate quantity management to water quality management, strengthening the relationship between water resources management and environmental management.

The framework is an instrument for the preservation of water body quality levels, which considers that human health and well-being as well as aquatic ecological balance should not be affected as a result of deterioration of water quality. It also considers that the costs of pollution control may be better suited when quality levels, evaluated by specific parameters and indicators, ensure their preponderant uses.

Table 2.3 shows the classification of fresh water, according to its prevailing uses, in five classes, as stablished in Resolution No. 357 of CONAMA. For each class, quality limits and/or conditions are established to be respected to ensure their preponderant uses, and the more restrictive, the more noble the intended use.

The effective instruments for environmental legislation are dependent on the framework of water bodies, which provide a clear vision of the desired uses and the quality of the water to be maintained. Although in some states this stage is already advanced, in most of the country, the bodies of water continue with the provisional classification (Class 2). Legislation should not be viewed as a simple benchmark of values but as goals to be achieved within a timeframe to be defined among polluters, environmental agencies, and river basin committees (Von Sperling 2001).

The framework of bodies of water is not necessarily based on their current state but on the quality levels that the waters should possess to meet the needs of the community. It is a pact established by the society that makes possible the compatibility between the management of water resources and environmental management, promoting the protection and recovery of water resources. The framework should preferably be included in the water resources plans, the result of a planning process that establishes the priorities of uses of the water bodies.

According to the PNRH proposal, it will be the responsibility of the water agencies or basin agencies, in the scope of their area of activity, to propose the framework of water bodies to the respective river basin committees, the National Council for the Environment (CNRH) or their State Water Resources Council (Conselho Estadual de Recursos Hídricos (CERH)).

Table 2.3 Classification of fresh water according to their prevailing uses

Classification of fresh water	Prevailing uses	Color
Special class	- Supply for human consumption, with disinfection - Preservation of the natural balance of aquatic communities - Preservation of aquatic environments in integral protection conservation units.	Blue
Class 1	- Supply for human consumption after simplified treatment - Protection of aquatic communities - Recreation of primary contact, such as swimming, water skiing, and diving, according to CONAMA Resolution No. 274/2000 - Irrigation of vegetables that are consumed raw and fruits that develop close to the soil and that are eaten raw without peel removal - Protection of aquatic communities in indigenous lands.	Green
Class 2	- Supply for human consumption, after conventional treatment; - Protection of aquatic communities; - Recreation of primary contact, such as swimming, water skiing and diving, according to CONAMA Resolution No. 274/2000; - Irrigation of vegetables and fruit plants and parks, gardens, and sports and leisure fields, with which the public can come into direct contact - Aquaculture and fishing activity	Yellow
Class 3	- Supply for human consumption, after conventional or advanced treatment - Irrigation of tree, cereal, and forage crops - Amateur fishing - Secondary contact recreation - The watering of animals	Orange
Class 4	- Navigation - The landscape harmony	Red

Source: Adapted from CONAMA Resolution No. 357/05

2.2.1.2 Grant of Right to Use Water Resources

A grant is an administrative act of authorization or concession in which the granting public authority allows the grantee to use the water resource, for a determined period, under the terms and conditions expressed in the respective act. By the Federal Constitution of 1988, the water is owned by the union or the states, having been established federal rivers and state rivers.

According to the Federal Constitution, all groundwater is state-owned, and therefore the National Water Agency does not authorize the use of groundwater in Brazil but rather the management bodies of the different states and the federal district. In practice, this procedure ends up being questionable, since the limits of the aquifers do not coincide with the limits of the states, and it is common to observe completely different groundwater use rules in bordering states, which are an inconsistency especially in cases wherein the different states refer to the same aquifer system.

The grant of the right to use water resources is probably one of the most important IWRM instruments in Brazil, as it is the distribution of available water resources among the different users, who eventually compete for scarce resources, in quantity or quality, to meet the needs. The grant must guarantee the user the right to use the water, conditioned to water availability. It should be noted, however, that this premise will only be achieved in those states that have a reliable information base on water availability in the river basins, allowing the grant to be made taking into account consistent information regarding the region's hydrological regime.

Law 9,433/1997 reserves the possibility of suspension or cancellation of grants in situations of extreme events, such as severe droughts, and prioritization of the uses of human supplies and animal watering. However, if the flows granted cannot be supplied to the users even in normal periods without extreme events, the situation becomes more complex and should be the object of ample reflection at the national level, as this may decrease the credibility of the system and increase the potential for conflicts among users.

The quantity to be granted varies with the hydrological regime of the river and according to the legislation, as explained in Fig. 2.2. In permanent or perennial rivers, the grant is usually made based on $Q_{7,10}$ or $Q_{90\%}$ or $Q_{95\%}$, and only part of the minimum flow values are granted.

In the case of temporary or intermittent rivers, the grant process becomes more complex, since in the dry season, the river ceases to flow and the values of $Q_{90\%}$ and $Q_{95\%}$ can be zero in cases where the watercourses cease to flow in natural regime for more than 5–10% of the time. In these cases, it is necessary to regulate the watercourses for grant purposes in the dry period of the year. For these streams it has been more common the use of regularized flows associated to different stays in time, as 90% (Q_{90reg}) or 95% (Q_{95reg}).

In the several Brazilian states, users have been requesting the respective water resources management agencies' permits for the abstraction of surface water and the exploitation of groundwater for diverse purposes, with irrigated agriculture being responsible for the largest number of requests for grant. The grant application processes are divided into requirements for surface or groundwater withdrawal, and in cases of surface water withdrawal, any interventions that change the watercourse quantitatively or qualitatively are included.

For each type of use, specific studies and information are requested. Normally, the required information includes the estimation of the minimum reference flow at the site of the enterprise to characterize the water availability. According to Law No. 9,433/1997, water derivations, abstractions, and storage of small population groups are considered insignificant but should be established within the scope of each state/ federal district or river basin.

2.2.1.3 Charging for the Use of Water Resources

Charging for the use of water has been foreseen in Brazil since 1934, with the promulgation of the Water Code by Federal Decree 24.643 of July 10, 1934. The Water Code incorporated modern concepts that remain advanced and current until today, as the charging. This code establishes that the common use of the waters can be free or paid, according to the laws and regulations of the administrative district to which they belong, establishing also that the works for water quality will be executed at the expense of the violators that, besides the criminal liability, if any, shall be liable for any loss or damage they cause and for fines imposed on them in administrative regulations.

The charging for the use of water was formally established in Brazil by Law 9,433/1997. It is foreseen to charging for the derivation of the water or for the introduction of effluents into the bodies of water, in view of their dilution, transportation, and assimilation, depending on the class of framing of the water body in question. The amounts collected from the use of water resources should be applied primarily in the river basin where they are generated.

According to ANA (2009), the costs for the use of water aim to recognize water as an economic good and give the user an indication of its real value, as well as encouraging the rationalization of water use and obtaining financial resources for the financing of the programs and interventions contemplated in the water resources plan of the river basin.

In the Brazilian model, the river basin committees will have the attribution of defining the values of the tariffs, based on unit prices and established maximum and minimum limits. The limit values will be established by the National Council of Water Resources (CNRH), in the case of water bodies of the union domain, or by the State Water Resources Council (CERH), for waters under state control. The resources will be applied on a participatory, decentralized, and integrated basis, according to the water resources plan for each river basin, with the purpose of avoiding waste and promoting the treatment and proper use of water, as well as the quality of the water and the environment, as recommended by the IWRM framework.

The agencies and entities managing water resources should prepare technical studies to subsidize the proposal of the amounts to be charged for the use of water resources, based on the mechanisms and quantities suggested by the Hydrographic Basin Committee to the respective Water Resources Council, as per clause VI of Art. 38 of Law No. 9,433 of 1997.

The river basin committees may institute incentive mechanisms and reduce the amount to be charged for the use of water resources, due to voluntary investments made by the user in studies, programs, projects, technologies, and actions to improve water quality and river regime, that result in environmental sustainability of the basin and that have been approved by the respective committee.

The charging for the use of water is an instrument of management of water resources, essentially economic, that acts in two senses: it promotes the control of the waste of water, provided that the user-payer is sufficiently burdened to take the necessary measures to streamlining its use (greater economic efficiency in the use of water), and generates financial resources for investments in programs for the improvement of water resources in the basin.

One of the necessary steps for the implementation of the user-pays principle is the creation of the Basin Agency, which should collect the tariffs for the use of water resources and the preparation of the investment plan of the financial resources collected, among other tasks. The committees have the power to deliberate on the resource allocation plan in the basin and its priorities. The Basin Agency is also responsible for financial support for programs approved by the committees and for the preparation of studies and reports on the situation of river basins. The Basin Agency is, therefore, the executing agency, the operational arm of the committees.

2.2.1.4 Water Resources Information System

According to federal legislation, basic principles for the functioning of the Water Resources Information System (Sistema de Informação em Recursos Hídricos, www.snirh.gov.br) are decentralization of the acquisition and production of data and information, the unified coordination of the system, and access to data and information guaranteed to society. In accordance with federal law, similar state laws, which institute the National Water Resources Policy (PNRH), also provide for the information system as a strategic instrument for the management and planning of water resources in state domains.

At the same time, the needs of the federal and state management bodies for the management of grant applications highlight the demand for tools based on geographic information systems (GIS) and relational database management systems capable of storing, processing, and making available information on the state of water resources.

According to the Secretariat of Water Resources (SRH) of the Ministry of the Environment, currently SRHU, for safe and effective management of the granting instrument, it is necessary to be aware of the water availability of the source; to know the present and future demands of the basin, so that the necessary water balance is realized and evaluated, as far as the acceptability and the quantitative and qualitative interferences desired and/or existing; to consider the guidelines for grant (grant criteria, framework, restrictions on use, etc.); and to adopt technical and administrative systematics for information processing and assessment of grant suits (MMA/SRH 2006).

In this case, a Web access information system can be conceived as a tool that integrates the storage and dissemination of innumerable information about water resources, produced and elaborated by several research and management entities, and associates with this database as rule-based administrative architecture capable of analyzing and managing the granting suits, keeping the river basin development scenario permanently updated.

2.3 Conclusions

In spite of advances in the management of water resources in Brazil after the promulgation of the Water Law, there is still a lot to be done, because despite having more than 20 years of its implementation, the instruments of IWRM in the aforementioned law have not yet been adequately implemented to balance the various problems existing in the country, both in quantitative and qualitative terms.

In relation to the grant of water use, it is verified that in many Brazilian states the grant is still based on scarce and/or inconsistent hydrological information, generating serious reliability problems in relation to the volumes granted to the different users.

Regarding the framework of water bodies, most of the Brazilian river basins have not yet implemented this important management tool, and, therefore, the stream courses are classified as Class 2, according to the legislation. This reflects the prioritization that has been given only to the quantitative aspects when issuing water grants, without considering water quality.

Concerning the charge for the use of water, it has been shown that the fee has not acted adequately to promote the efficient use and control of water waste and that the resource generated by the collection has not yet resulted in significant improvements both at the point of quantitative and qualitative view of water at the river basin level.

References

ANA (2009) Água: Fatos E Tendências Brasília. Agência Nacional De Águas, Brasília

ANA (2013) Conjuntura dos recursos hídricos no Brasil: 2013. Agência Nacional De Águas

ANA (2015a) Conjuntura dos recursos hídricos no Brasil: Informe 2014. Agência Nacional De Águas, Brasília

ANA (2015b) Encarte especial sobre a crise hídrica. Agência Nacional De Águas, Brasília

Brasil (1997) Política Nacional de Recursos Hídricos (PNRH) vol Lei no 9.433

ICWE (International Conference on Water band the Environment) (1992) International Conference on Water and the Environment, The Dublin Statement on Water and Sustainable Development. Geneva: ICWE Secretariat.

Kemper KE, Blomquist WA, Dinar A (2007) Integrated river basin management through decentralization. Springer, Berlin

MMA/SRH – Ministério do Meio Ambiente. Secretaria de Recursos Hídricos. Estágio atual da política nacional. Disponível em: http://www.mma.gov.br/port/srh/index.cfm. Acesso em: 10 jul. 2006.

United Nations (1992) Sustainable Development e United Nations Conference on Environment & Development Rio de Janeiro, Brazil, 3 to 14 June 1992, Agenda 21. https://sustainabledevelopment.un.org/content/documents/Agenda21.pdf. (Accessed 9 March 2018).

UN (2017) O Direito Humano à Água e Saneamento. Comunicado. 8

Von Sperling M (2001) A legislação ambiental e o controle da poluição das águas. In: Anais do I Encontro de preservação de mananciais da zona da mata mineira. Viçosa-MG: ABES, ABAS, UFV. 2001, p.34–35.

Chapter 3
The Necessity of IWRM: The Case of San Francisco River Water Conflicts

Valmir de Albuquerque Pedrosa

Abstract The search for agreements to decrease water conflicts requires to understand the clear nature of the dispute. The best way to explain it would be using a real example of a Brazilian river which crosses several states and whose use of water had nationwide consequences. It would be good that practically all possible uses of the water were included – sanitation services for the cities, irrigation, industry, mining, power generation, tourism, artisanal fishing and fish farming and preservation of river mouth ecosystem, among others, including the transposition of its waters. Regarding the amount of water offered, it would be interesting to analyse a river in which a recent scarcity made it possible to assess the effectiveness of the response to this scenario. It would also be enrichening to choose a river with a relevant amount of water network infrastructure installed, as well as the presence of all instruments of the National Policy and all the parties forming the National Water Resources Management System. Besides that, it would be educational to detail a case with unquestionable need for an integrated water resources management regarding surface water, groundwater and costal water.

Keywords Multiple use of water · Energy power plant · Water use for irrigation

3.1 Introduction

The example of IWRM is presented in the case of the São Francisco River – the river of national unity in Brazil. This river has strong historical components, an old and expressive cultural richness and artistic traditions and the exuberant presence of several biomes – the Cerrado, the Caatinga and the Atlantic forest – and ecosystems, estuaries, lagoons, lakes, veredas (creek bed) and others.

V. A. Pedrosa (✉)
Federal University of Alagoas, Maceió, Alagoas, Brazil
e-mail: valmirpedrosa@ctec.ufal.br

E. O. Vieira et al. (eds.), *Integrated Water Resource Management*,
https://doi.org/10.1007/978-3-030-16565-9_3

For this reason, the case of the São Francisco River was chosen to perfectly exemplify the example of IWRM in the nature of the water conflict – the central theme of this chapter. The notable facts regarding these waters are ancient. Serious water conflicts have arisen in the São Francisco River basin. Considering the demand, there has been increasing consumption of water to serve the urban areas, promote irrigated agricultural activities, ensure power generation, guarantee navigation, boost tourism, develop fishing, and preserve the fauna and flora, among others. Considering the offer, there has been a progressive unavailability of water because of pollution; there has been degradation of river springs and margins and increasing deforestation, all that contributing to diminishing average discharge availability for use.

The worldwide debated phenomenon of global warming can also be another threat, broadening water conflicts by intensifying and increasing the frequency of water extremes such as prolonged droughts.

In the CBHSF (São Francisco River Basin Committee) webpage, it is written that the "São Francisco river basin encompasses a catchment area of 639,219 km^2 (7.5% of the country) and an average discharge of 2850 m^3/s (2% of the country total). The São Francisco river is 2700 km long and its spring is located at the Canastra Mountains in Minas Gerais, the river flows northwards through Bahia and Pernambuco where it alters its course eastwards reaching the Atlantic Ocean through the border of Alagoas and Sergipe. The basin encompasses seven states – Bahia (48.2%), Minas Gerais (36.8%), Pernambuco (10.9%), Alagoas (2.2%), Sergipe (1.2%), Goiás (0.5%), and the Distrito Federal (0.2%) – and 507 municipalities (approximately 9% of all municipalities in the country)" (CBHSF 2016).

Despite its average discharge, because of the severe drought of the last 5 years, discharge at the river mouth in January 2017 was at 700 m^3/s. This level of discharge paralysed commercial navigation in some stretches of the river, reduced power generation, diminished effective irrigation area in certain parts of the basin and made it possible for saltwater intrusion to reach the town of Piaçabuçu, in Alagoas. Besides that, the point of tapping river water for sanitation companies was altered to continue the water supply of cities, among other effects.

To exemplify the severity of water conflict and irrigated agriculture expansion, according to Bahia Irrigation Association website (AIBA 2016), in June 2016, the farmers "decided to suspend irrigation in more than fifty percent of the irrigated area because they were concerned about the low discharge of the rivers supplying the west of Bahia. This means that 72 thousand out of 120 thousand of irrigated hectares had their irrigation equipment turned off" (AIBA 2016).

During the same crisis, the Nilo Coelho Irrigation District located in Petrolina (PE), 23 thousand irrigated hectares, informed all users in September 2015 that "because of the forecast of more severe low levels at the Sobradinho Dam, it would adopt an operation regime of rationing starting on October 26, 2015 when Sobradinho would reach a 5.14% volume".

In September 2016, the CBHSF approved the São Francisco River Basin Water Resources 10-year plan for the periods 2016–2025 with the support of the AGB Peixe Vivo (Peixe Vivo River Basin Management Support Executive Association).

Table 3.1 Withdraw demands by user sector (m^3/s) in the São Francisco River

User sector	2000	2006	2010
Human supply	26	27.3	31.3
Rural supply	3.8	3.7	3.7
Irrigation	114	123.3	244.4 (2013)
Livestock	6.7	9.1	10.2
Industrial supply	15.3	17.4	19.8
Total	165.8	180.8	309.4

Table 3.2 Withdraw discharge (m^3/s) by consumptive use in the São Francisco River

User sector	Total	Surface	Groundwater
Human supply	31.31	27.18	4.12
Rural supply	3.71	0	3.71
Irrigation	244.38	233.83	10.55
Livestock	10.10	1.86	9.02
Industrial supply	19.81	15.59	4.22
Total	309.44	277.80	31.64

Taken from the plan, Tables 3.1 and 3.2 present the evolution of water demands and its distribution per user sector (CBHSF 2015). In Table 3.1 it is possible to see that withdrawals for the irrigation sector jumped from 114 m^3/s in 2000 to 244.4 in 2013. In general, offtake jumped from 165.8 m^3/s in 2000 to 309.4 m^3/s in 2010, an 87% increase over a 10-year period.

Table 3.2 shows offtake discharge divided by user sector and surface or groundwater source. The expressive participation of irrigation over water demands is a repeated fact, with slight variations, in countries that have an agricultural vocation. It is possible to see that withdraws from groundwater represent about 10%.

In times of water crisis as the one taking place in the São Francisco River (RSF), it is even more important to emphasize the indissoluble relationship between surface waters and groundwaters. In the case of the São Francisco River, this example can be highlighted by the Urucuia aquifer. According to studies of the National Water Agency "the average contribution of Urucuia to form the RSF baseflow in a period of drought is 80% using the Sobradinho Dam as reference. The Urucuia system practically maintain the São Francisco river during a drought period. The Bambuí aquifer is also responsible for the formation of the discharges in periods of no rain, forming the discharges of the Grande River as well as being the source of water for the town of Sete Lagoas" (Santana 2015). There is a consensus between the states of Bahia and Minas Gerais that there are too many authorized wells and that a more intense inspection is decisive to get to know the amount of water extracted from these aquifers so important to the São Francisco River basin.

The Urucuia aquifer is located on the west of Bahia where there are large irrigation projects in the cities of Barreiras, São Desidério and Luís Eduardo Magalhães.

The connection between surface waters and groundwaters is located in the area of the Grande, Corrente, and Carinhanha rivers where there are several cases of conflicts. The aquifer potential is high; there are wells that can supply up to 600 m^3 per hour, a discharge able to meet the need of a city with 90 thousand people with 150 litres per person per day.

It is easy to see the diversity of conflicts along the basin, especially regarding dam operations and other uses – supply to cities, irrigation, tourism, navigations, fishing and fish farming, mining, water quality, forestry, preservation of veredas, lagoons and estuary ecosystems.

In the report from the consultant Rosana Garjulli, which reported the Multiple Uses of São Francisco River Water Workshop, the conflicts were contextualized per stretch as follows (CBHSF 2013a):

1. On the streambed stretches located at the sub medium and low São Francisco, there are conflicts regarding the way the dams are operated to generate power, determined by the System National Operator (ONS) and managed by the CHESF with the other user sectors (human supply, navigation for passengers, loads and fish transport, irrigation, fishing, fish farming, dryland farming, tourism).
2. In the medium São Francisco, there is the indiscriminate use of waters (including groundwater) and the inadequate use of the soil through irrigation leading to conflicts with other uses.
3. In the high São Francisco, there is also conflict between the operation of the Três Marias dam for power generation managed by CEMIG but operating according the determinations of the ONS and the other uses (tourism, sports, leisure, supply, irrigation, navigation, fishing and fish farming).

Besides the above-mentioned conflicts, the report highlighted some gaps in the implementation of the water resources management system which should make provisions for multiple uses of water.

For example, in California, the large reservoirs and the distribution canals have always been approved after a critical period of drought. In 2015, going through a 5-year period of below-average rainfall, the government of California got the political support of the population to prohibit the washing of sidewalks, to allow garden watering only once a week, to compulsorily reduce urban consumption up to 25% (with fines for infractions), to seriously inspect the consumption for irrigation prohibiting non-authorized uses or to use above allowed limits, among other measures. The governors were pushed to take these severe measures by the critical scarcity of water. The same is true for Brazil; certain unpopular but necessary measures are only taken and carried out in times of emergency.

Going back to the case of the São Francisco River, although the offer-demand relationship is already fierce, all states in the basin have irrigation and city supply projects to be implemented, as, for example, is the case of the São Francisco River Integration Project (PISF), which will soon start full operation taking water from the São Francisco River to the states of Pernambuco, Paraíba, Rio Grande do Norte and Ceará. The offtake discharge will vary from 27.4 m^3/s to 127 m^3/s conditioned to the amount of water reserved in the Sobradinho Lake.

In the basin Water Resources Plan for the 2016–2025 period (CBHSF 2015), conflict situations have been summarized as follows:

> "The results of the water balance show that there are cases of over-exploration of available water resources and conflicts over the use of water. The main conflicts result from the difficulty to settle meeting water demands for consumptive uses with the requirements for power generation and the competition for water of the several consumptive uses, irrigation especially because of the water volume required.
>
> The water uses in the main stream of the São Francisco river basin are conditioned by the hydroelectrical power stations operations. The water volume used in the production of energy is several times higher than the volume required by other consumptive uses".

Water transport is also affected by water scarcity. Navigation in the São Francisco is as old as its history. However, recent drought years have had a significant impact on this type of transport. As the discharges of the São Francisco River decreased, navigation was jeopardized. Commercial transportation was completely halted in July 2014 (Carvalho 2014). The ICOFORT company specialized in the transport of cottonseed explained the situation as: "Our entire production will be transported by road, increasing the damage to the environment, bringing the need to invest in roadways, increasing the risk of accidents and the costs, which will raise final product price between 20% and 30%".

Water conflicts do not take place only in the mainstream of the river. There are several conflicts on the tributaries of the São Francisco. In the book *Velho Chico – A Experiência da Fiscalização Preventiva Integrada na Bahia* (Ministério Público do Estado da Bahia 2014), there are descriptions of four water conflicts in the rivers of Bahia, all of them tributary of the São Francisco River: (i) the case of the Salitre River, (ii) the case of the Mirorós Dam, (iii) the case of the town of Lapão and (iv) conflicts caused by the small hydroelectric power stations (PCH). These cases will be presented next, and they are a summary of the information contained in the book mentioned above supplied vegetables, fruits and greens to several cities.

The reason behind the conflict in the Salitre River is the interruption of the water course due to 35 dams. These reservoirs were built in the 1970s, before it was necessary to be granted authorization and before the need to have a forecast of bottom discharge to maintain a minimum ecological flow. In order to mitigate the situation, CODEVASF built nine embankment dams to make the low Salitre River perennial with the waters from the São Francisco River, thus making the agricultural activities of the Salitre Vail Associations Union possible. This collision of interests between the city and the agricultural sector led to scarcity of water for the cities. In 1970, the government of Bahia and the Juazeiro Town Hall tried to limit the conflict, allowing each family to irrigate a maximum of 3 hectares. Not everybody complied with the regulations, and as time went by, the conflict increased. In February 1984 it reached its peak with an armed confrontation, which resulted in deaths when the residents turned off the power grid feeding large offtakes. Since then, the situation only got worse.

In March 2010, the Salitre Project was implemented, aiming to irrigate 34,000 hectares. The project reserves 20% of the area for small farmers, each one with 6 hectares. The several requirements to get one of the plots left some of the residents out of the project, "aggravating even more the water conflict in the area" (Ministério Público do Estado da Bahia 2014).

In 2011 there were new violent conflicts; some light poles were turned down and wire cut to prevent the large irrigation pumps from working which also affected schools, houses and medical stations. The remaining Quilombola community Lages dos Negros was affected by the conflict, school classes were suspended and medical stations closed, the commerce collapsed and sometimes even the houses have no electricity. In October 2010, the Salitre River Water Basin Committee required that basin users be registered, which has not happened yet.

3.2 The Case of the Zabumbão

In the CBHSF newspaper, edition 33 of August 2015, there is an interview with the president of the Santo Onofre and Paramirim rivers' Water Basin Committee, right margin tributaries of the São Francisco River. The president said that the conflict that has called the attention of CBHSF takes place in the Zabumbão Dam located on Paramirim River basin and built by CODEVASF. The reservoir currently supplies water to four municipalities: Paramirim, Caturama, Botuporã and Tanque Novo (Caires 2015).

Recently the government of Bahia opened a public bid to build a new water supply system with water from the Zabumbão to supply six other municipalities in the area: Rio do Pires, Ibipitanga, Macaúbas, Oliveira dos Brejinhos, Boquira and Ibitiara. The new infrastructure will help the government increase offtake from 100 litres/s to 523.9 litres/s. The committee president affirmed that some municipalities do not need the water and that the water supply system "will dry off the Zabumbão", overly affecting water safety in the area. Recent droughts led to the reservoir volume not being able to go past half of its capacity (Caires 2015).

As an alternative to this project, the committee president defends modernization, and a more efficient use of the water is used in the Paramirim Vail irrigation, where flood irrigation is used in 1300 hectares, as well as the treatment of the sewage upstream the reservoir and the construction of two new dams in the Caixa and Remédios rivers (Caires 2015).

3.3 Conclusions

In conclusion, this article highlights several water conflicts in São Francisco River basin. To overcome these conflicts, the São Francisco River committee recommends that the National Water Agency (ANA) and the water management agencies of the states integrating the basin with the arbitration of the CBHSF and the Tributary Basins Committee celebrate an integrated management partnership as the starting point to build a water pact in the basin defining the rules for the sustainable use of its water resources.

The Multiple Uses of São Francisco River Water Workshop (CBHSF 2013a, b) recommended and reinforces the need to build this water pact. In the São Francisco River basin, this pact should at least:

1. Define and implement mechanisms to articulate and integrate actions of state, federal, CBHSF and CBH tributaries management agencies.
2. Promote and implement an effective campaign to regulate water use in the entire basin including the updating of registrations and concession authorizations.
3. Establish criteria for a process to review granted authorizations assessing and redefining them according to effective water use capacity and availability.
4. Define strategies to strengthen the regional consultation chambers aiming at more dialogue with the basin society.
5. Structure an integrated inspection system at federal and state levels articulated with the environmental system.
6. Identify dialogue channels, and define articulation strategies for CBHSF with regional leaderships and collegiate of other public policies (identity territories, EPA management councils, municipal health and education councils, mayor and alderman associations, public ministry, and others) aiming to ensure more integrated action in the basin.
7. Encourage the participation of different CBHSF instances in the municipal and state basic sanitation plan preparation process aiming to get them closer to the priorities identified in the São Francisco River Basin Plan.
8. Articulate with the environmental agencies the demarcation and recovery of the Permanent Protection Area of the Sobradinho Lake and downstream the dam.
9. Promote an extensive and integrated water quality monitoring program (saltwater intrusion, phosphorus, others).
10. Review the charging criteria and rates for water use according to user size and the "polluter pays" principle.
11. Create a work group to deepen knowledge, and propose the proper management of the intermittent rivers in the basin.
12. Promote the articulation between the Tocantins and São Francisco River basins to discuss possible integration of the basins.
13. Identify articulation mechanisms in the federal, state and municipal instances to make viable the planning and implementation of basic sanitation actions in an integrated and sustainable way.

References

AIBA (2016) Agricultores do Oeste da Bahia reduzem mais da metade da área irrigada devido à escassez hídrica. Associação de Agricultores e Irrigantes da Bahia. http://aiba.org.br/noticias/agricultores-do-oeste-da-bahia-reduzem-mais-da-metade-da-area-irrigada-devido-escassez-hidrica/#.W4mowC2ZNTY

Caires AB (2015) Barragem vira polêmica na Bahia vol 33

Carvalho D (2014) Seca atrapalha navegação do Rio São Francisco. São Paulo

CBHSF (2013a) Oficinas “usos múltiplos das águas do Rio São Francisco”
CBHSF (2013b) Usos múltiplos na bacia hidrográfica do Rio São Francisco: política e prioridades
CBHSF (2015) Minuta de plano de recursos hídricos da Bacia Hidrográfica do Rio São Francisco 2016–2025, RP1A diagnóstico da dimensão técnica e institucional
CBHSF (2016) ABACIA: principais características. Comitê da Bacia Hidrográfica do Rio São Francisco. http://cbhsaofrancisco.org.br/2017/a-bacia/. Accessed July 2016
Ministério Público do Estado da Bahia (2014) Velho Chico: a experiência da fiscalização preventiva integrada na Bahia
Santana A (2015) Vida que brota debaixo do barro do chão. Comitê Da Bacia Hidrográfica Do Rio São Francisco, Belo Horizonte

Chapter 4
Water Resources Management in California

Samuel Sandoval-Solis

Abstract California has an intense history of water management and resources manipulation. The main drivers for some of the largest water management infrastructure projects are (1) a spatial mismatch between where most of the precipitation falls on the state and where most of the water is needed and (2) a temporal mismatch of precipitation during winter months and the agriculture season on summer. This chapter describes the legal framework and water allocation systems to manage surface water, groundwater, and environmental water that are guiding California toward adopting an integrated water resources management framework.

Keywords Groundwater management · Water rights · Water allocation · SGMA

4.1 Background

Depending on the economic cycles, California fluctuates between the 5th to 7th largest economies in the world (Corcoran 2018; Forbes 2017). Furthermore, there is no other state in the United States (USA) wherein the economic development has been linked so tight to the water resources development, for instance, in 1849 during the gold rush when miners used high-pressure jets of water to wash entire hillsides (a procedure called hydraulic mining) to excavate sediments and separate the gold at industrial proportions. Nowadays, water still supports the economic development of the state such as providing clean pure water from Hetch Hetchy aqueduct to produce computer processors in Silicon Valley, all the way to collecting and transporting rainfall and snowfall originated in the Northern California, Wyoming, Colorado, and Utah into Southern California to support the entertainment and movie industry.

S. Sandoval-Solis (✉)
Department of Land, Air and Water Resources, University of California, Davis, Davis, CA, USA
e-mail: samsandoval@ucdavis.edu

E. O. Vieira et al. (eds.), *Integrated Water Resource Management*,
https://doi.org/10.1007/978-3-030-16565-9_4

Because water is not equitably distributed in time and place, in the right quantity with the adequate quality, a discipline called water resources planning and management (WRPM) is used to redistribute the resource in a way that satisfies the needs of water users, including the environment, today and in the future (Loucks et al. 2005). In California, WRPM has been widely used because there is a mismatch between the hydrologic cycle and when and where water is needed. First, the Mediterranean climate brings precipitation (rainfall and snowfall) during winter and snowmelt during spring. However, water is needed year-round for cities and the environment, during summer for agriculture and during certain hours of the day in summer to supply electric energy at peak hours (Hanak et al. 2011). To close this temporal gap, an important amount of natural water storage (such as aquifers) are managed, and man-made reservoirs (such as dams) were built throughout the state (USBR 2018). The main purpose of a reservoir is to store water during seasons of water abundance to release it later when water is needed. Some reservoirs have such a large storage capacity that can store water between years.

Similarly, WRPM is widely used in the state of California because there is a spatial mismatch between where precipitation occurs and where water is needed (Fig. 4.1). Generally, precipitation occurs in the northern part of the state and along the Sierra Nevada. In contrast, water is needed everywhere in the state, but mostly (a) along the coast, i.e., the San Francisco Bay Area; (b) in the center of the state, i.e., the entire Central Valley (Sacramento, San Joaquin, and Tulare basins) for agriculture production; and (c) in in the south, in Southern California to provide water to three quarters of the population of the state and the Imperial Valley. To close this spatial gap, an important amount of man-made conveyance infrastructures (e.g., canals, aqueducts, tunnels, and pipes) were built throughout the state (DWR 2018). The main purpose of this infrastructure is to transport water where it is needed; the California Aqueduct and the Delta-Mendota Canal move water from the north to the south of the state; the Hetch Hetchy and Mokelumne aqueducts transport water from the Sierra Nevada to the San Francisco Bay Area; Los Angeles and Colorado aqueducts move water to Southern California to supply urban and agriculture water needs.

Integrated water resources management (IWRM) is a process which promotes the coordinated development and management of water, land, and related resources, in order to maximize the resultant economic and social welfare in an equitable manner without compromising the sustainability of vital ecosystems (GWP 2000). Unfortunately, in California, IWRM has not been applied until the last decade. During the nineteenth and twentieth century, water supplies, infrastructure, and economic and land resources were planned and managed independently, without much of coordination. Many of the water projects were designed and built independently, or in a piecemeal fashion as one piece of a bigger system, or they were developed and built (or not build) opportunistically when there was enough political willingness and resources. For instance, there is no single water code that explains the rules of how water is used throughout the state. In lieu, there is a body of legal cases and legislation that have been passed throughout the years describing the procedures,

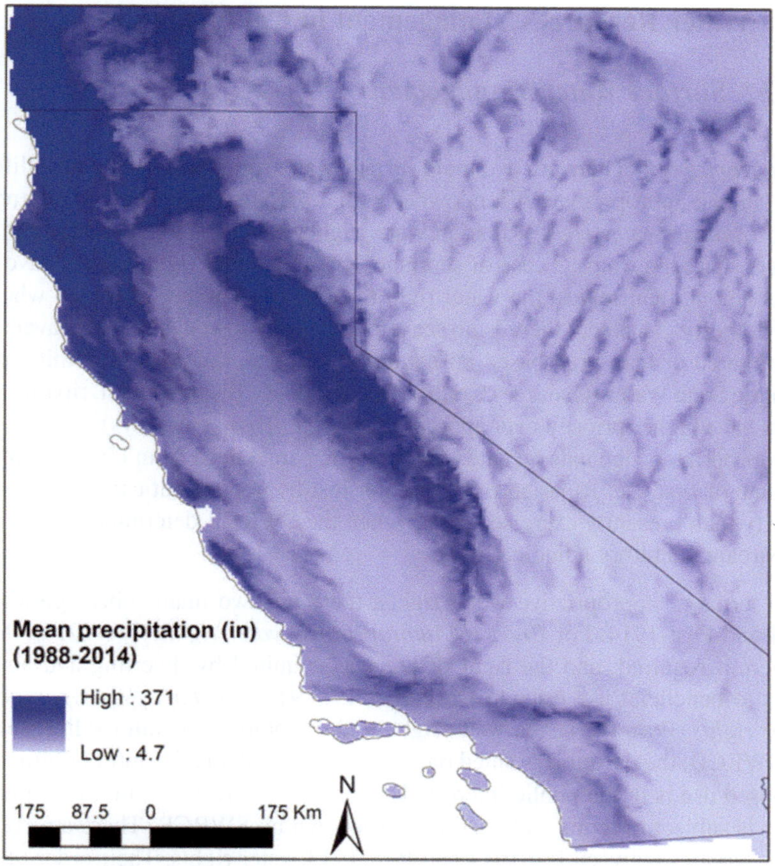

Fig. 4.1 Mean precipitation in the state of California

priorities, and allocation policies to distribute water in the state. In addition, there was no protection to the environment until the California Environmental Quality Act (CEQA) (Olshansky 1996), the Clean Water Act (2008), the Endangered Species Act (Congress 1973), the designation of wild and scenic rivers (Tarlock and Tippy 1969), and recently all the lawsuits to supply environmental flows to the mainstem of the most important river in the state. The only common rule is that water in the state of California must be used reasonably and for a well-defined beneficial use (Trelease 1957).

Depending on the *water source*, there are different *legal frameworks*, *decision-making processes*, and *water allocation systems*. Unfortunately, this means that water resources management in California is a fragmented system (opposite to integrated) that depends on the specific context. The following section describes these processes.

4.2 Water Resources Management in California

4.2.1 Surface Water Management

Legal framework There are two main surface water rights in the state of California: riparian and appropriative. A landowner with a property adjacent to a watercourse has a *riparian water right*. Riparian water rights are correlative, meaning that the owners share the water in case of shortage. Riparian water right holders have equal rights among themselves; the water right must be used in the watershed where the land property is located; there is no seasonal storage allowed. The water availability is estimated or measured based on natural conditions; there is no permit required from the State Water Resources Control Board (SWRCB). In general, riparian water rights have higher priority over appropriative rights (but not always); priority vis-à-vis appropriators depends on date of patent (the parcel deed from US Government), not date of first use. Riparian rights are not lost by nonuse but can be given lower priority than presently exercised rights when the SWRCB determines all the rights to a stream (statutory adjudication).

In terms of appropriative water rights, there are two main subcategories: Pre-1914 and Post-1914. For *Pre-1914 appropriative water rights* (Pre-1914), there is no permit required, and the right of use was acquired by diverting and applying water to beneficial use prior to December 19, 1914. For *Post-1914 appropriative water rights* (Post-1914), the water right holder requires a permit (or license) from the SWRCB; the permit is granted only if water is available for appropriation and if proposed use is in the public interest. The amount of diversion, timing, and use of water is subject to terms and conditions specified by SWRCB. There are common characteristics between Pre-1914 and Post-1914 water rights. Their order of water allocation, commonly known as priority, is based on time of use (for Pre-1914) or date of application (for Post-1914); this prioritization is usually referred as first in time, first in right. In times of scarcity, later (junior) appropriators are cut off before earlier (senior) appropriators. That is, early priority rights must be satisfied before later rights receive any water. The water right has a defined amount of water, which is the maximum amount of water that can be diverted under that right. However, this amount is not necessarily available in every year, and it can be cut back during drought periods. The water right title specifies the type and place of use and point and period of diversion. Furthermore, a water right may be lost through 5 or more years of nonuse (this action is commonly referred as use it or lose it) (Littleworth and Garner 2007).

Water allocation system Typically, riparian water rights are assigned first than appropriative water rights, this is because riparian water rights withdraw water during the wet season (winter and spring, when water is naturally available) and the amount of water withdrawn compared with the appropriative water rights is smaller. First, the natural flow is estimated for every basin and at different point of interest, called control points. Control points are usually located at streamflow gauges.

Second, the water available for every control point is estimated by subtracting the environmental flow requirements (see Environmental Flows section) to the natural flow. Third, the available water is compared with the riparian water right volume associated at a specific control point. Finally, the SWRCB decides if there is enough water available to supply the entire volume of water compromised in the water right permits or a portion of the total volume in which case the shortages are distributed equally by calculating the share among all riparian water users. In practice, the SWRCB monitors the water diversions and compliance of environmental flows by monitoring control point located at streamflow gauges. At the end of the water year, September 30, every water right holder submits a report to the SWRCB notifying the amount of water diverted, so the authority has an estimate of the water diverted.

As mentioned before, because of the Mediterranean climate of California, riparian water right holder mostly withdraws water during winter and spring. One of the main technical problems associated with riparian water rights is the estimation of the natural flows. In control points located at the headwaters with no major infrastructure upstream (such as weirs, dams, canals, tunnel intakes, major land use change, etc.), the natural flow is considered the streamflow at a determined gauge station. Downstream of major infrastructure, the natural flow is determined by using a mass balance method to calculate the naturalized streamflow (Wurbs 2006) or water resources modeling to determine the unimpaired flows (Kadir and Huang 2015).

Appropriative water rights have a different allocation system. Appropriative water rights have usually the following two characteristics: (1) water for appropriative water rights are usually stored in surface water reservoirs and transported through natural (mainstem of the river) and man-made infrastructure (canals and pipes), and (2) water can be stored and released at the discretion of the water right holder, meaning the period of use can be all year around (municipal users) or during specific portion of the years (e.g., the growing season for agriculture users). First, every April 1, the available water for appropriative water rights from every reservoir is estimated by subtracting the environmental flow requirements, conveyance, and evaporation losses to the reservoir storage on that date. Second, the water available for every reservoir is allocated in an orderly manner based on the priority (seniority) of the appropriative water rights (first in time, first in right). Third, in case of drought, junior water right holders can be notified that they can withdraw only a portion of their water right or no water at all. This procedure is performed by the institution or agency that owns and operates a determined reservoir (e.g., the US Bureau of Reclamation for the Central Valley Project or the California Department of Water Resources for the State Water Project, Sonoma Water for Coyote Valley Dam, etc.), and the allocation plan is submitted to the SWRCB for approval. In case of basins with no reservoirs, water right holders only can withdraw water after the flow at determined control points (usually streamflow gauges) are equal or above certain flow threshold. These thresholds have been previously calculated by the SWRCB to guarantee that environmental flows and senior water right holders will be able to withdraw water before junior water right holders.

4.2.2 Environmental Water Management

Legal framework There are three legal frameworks under which environmental flows, water to support the aquatic and riparian ecosystems, are determined and implemented in the state of California: (1) the Endangered Species Act (ESA) (Congress 1973), (2) Section 5937 of Fish and Game Code (Bork et al. 2011), (3) the Federal Energy Regulation Commission (FERC) relicensing process for hydropower dams, and (4) public trust doctrine (Sax 1970).

The ESA has been used historically as a mechanism to sue the owners and operators of water resources infrastructure to modify their operations, so they are not detrimental of endangered species. Usually, a nongovernmental organization (NGO) or an environmental agency within the government (e.g., US Fish and Game) is the institution suing the infrastructure-operational agency to modify their operations, mostly releases from reservoir. During the legal process, the court request biological opinions (BOs) to determine what are the impacts of the current reservoir operation as well as potential alternative strategies to support endangered species. The BOs are used as the basis and guidelines to modify reservoir operations. This strategy has been used in the mainstem of the most important rivers in California, because reservoirs were built in these places.

Section 5937 of Fish and Game Code states that "the owner of any dam shall allow sufficient water at all times to pass over, around or through the dam, to keep in good condition any fish that may be planted or exist below the dam" (Bork et al. 2011). Thus Section 5937 enforces the longitudinal connectivity of rivers by allowing fish to migrate upstream and downstream. Similar to ESA, there should be an institution (usually the California Department of Fish and Wildlife) that request infrastructure-operational agency to modify their infrastructure for allowing the passage of fish up and down the reservoir.

The FERC relicensing process for hydropower dam forces the hydropower owners and operators (HP-operator) to renew their dam operation license every determined number of years (usually between 20 and 30 years). When the HP-operator is renewing their license, FERC request that the proposed future operations are protective of any endangered species and in general of the environment. Thus, HP-operators are forced to design dam operations that meet state regulation (Section 5937) and that are protective of the environment and economically feasible to continue producing electricity. This procedure occurs mostly in relatively small dams built in the headwaters of different rivers.

The public trust doctrine is a state power that allows an institution (the SWRCB) to protect natural resources, including water, for the common benefit of the public above any individual benefit. The public trust doctrine allows the state of California to preserve and conserve rivers for the benefit of the public above any individual benefit. This policy is commonly used to protect or restore natural resources along the riparian corridor, including rivers, estuaries, and other water bodies for the entertainment of the public today and in the future. The SWRCB use this doctrine in degraded watershed where overuse of water has led to ecosystem degradation. The public thrust doctrine was applied in different legal cases; it is most frequently used to protect or restore tributaries of main rivers when the aquatic and riparian ecosystems living on the tributaries have been degraded. In addition to the previous legal mechanisms, in 1972 the state of California enacted the Wild and Scenic Rivers Act that protects many rivers throughout the state, including the Eel and Klamath rivers.

Water allocation system Typically, environmental flows have the highest priority, and water is allocated first to this water use because (a) environmental flows were determined out of a federal regulation (ESA or FERC relicensing) or (b) it is a state of California mandate to implement a regulation (Section 5937) or to protect natural resources (public trust doctrine) for the benefit of the public. Environmental flows, however, have been mostly misunderstood. While about 50% of the total water on the state is considered to be allocated to the environment, the percentage of allocation is not evenly distributed across the state. Most of the environmental water is allocated in the North Coast rivers where there is little competition with other users (PPIC 2016).

4.2.3 Groundwater Management

Legal framework Before 2014, there was groundwater management in the state of California; however, (a) it was localized in certain basins, mostly in Southern California through groundwater basin adjudications, a legal, contentious, and costly procedure to allocate the water of an aquifer; or (b) it was voluntary; local water agencies developed and implemented groundwater plans as a requirement to pursue economic incentives from the state of California in the form of bonds. These actions were not statewide; thus, benefits were only local. For instance, in certain basins of the state of California, there are still long-lasting problems of groundwater overdraft, land subsidence, and seawater intrusion (Zektser et al. 2005).

The most recent 5-year drought (2011–2016) changed the previous legal framework; the Sustainable Groundwater Management Act (SGMA), passed in September 2014, is the first legislation in the state of California to manage groundwater resources. SGMA does not provide a definition for sustainability; however, it does describe a list of undesirable results that must be avoided to achieve sustainable groundwater management. Each groundwater basin should be managed to avoid any of the following undesirable results: (1) chronic lowering down of groundwater table, (2) groundwater overdraft, (3) land subsidence product of groundwater overdraft, (4) seawater intrusion, (5) recharge of degraded water into the aquifers, and (6) adverse impacts on beneficial uses of surface water due to groundwater operations. SGMA considers three main steps for its implementation to make sure that none of these undesirable results occur in any groundwater basin. First, groundwater sustainable agencies (GSAs) were created for implementing SGMA. The legislation incentivizes that GSAs are formed by members of local organizations that are already managing water (such as irrigation districts) or land (county government, cities) on the overlaying area of the groundwater basin. Members from other groups of interest (e.g., tribes, agriculture, and environment) can be included in the GSA organization, if the members first listed agree to include them as member of the GSA. Conversely, if the local institutions do not agree to form a GSA, then the SWRCB will step in and implement SGMA. GSAs are allowed to keep records of wells, install water meters for monitoring water use, and impose fees for groundwater extraction and if necessary a moratorium on groundwater extraction if undesirable conditions persist. At this moment, more than 100 water agencies have been formed throughout the state of California. Second, water budgets (WB) for every water source will be estimated to provide a current diagnostic of each groundwater basin. A mass balance for every water source available in basin (surface water, groundwater, recycled water, desalinated water) will be estimated to determine the water supply, water use, and change of storage. This diagnostic will help to identify if the basin is currently experiencing any of the undesirable results. Third, groundwater sustainable plans (GSPs) will be developed to identify strategies that will impede or prevent any of the undesirable effects for happening by 2040.

Water allocation system For the adjudicated basins, the adjudication verdict prescribes the amount of water that each water user can extract from the ground, in which order and when. For the groundwater basins managed through SGMA, the groundwater allocation system will be defined for each groundwater basin by the respective GSP. Each GSP will manage groundwater basins by dividing it into subregions. Within each subregion, the overall groundwater recharge and extractions will be determined. Every groundwater user will have a maximum quota for groundwater extraction determined on the subregion's water balance. GSAs will monitor each groundwater user by installing water meters on wells. Also, GSAs will collect fees to manage and monitor each groundwater basin. Groundwater recharge and extractions will be monitored and managed to avoid any of the six undesirable results. Each GSP will include a suit of strategies to manage: (a) water demands that rely on groundwater resources, such as water conservation strategies for cities to reduce water use, incentives for changing agricultural production to less water-intensive crops, deficit irrigation techniques, and land use fallowing, just to mention a few strategies, and (b) water sources to increase aquifer recharge through active or passive managed aquifer recharge (MAR). Active MAR includes the construction and operation of recharge basins that divert excess surface water during the rainy season to infiltration ponds where water infiltrates into the aquifer. Passive MAR strategies include in lieu groundwater banking where groundwater users temporally use surface water during years where there is an excess of surface water reservoirs letting the aquifer rest and actively use groundwater when surface water resources are scarce.

4.3 Looking into the Future

SGMA is highly influencing water resources management in the state of California. For the first time in the history of water resources management in the state, there is a legal clause in a regulation that explicitly protects the interaction of surface water and groundwater. This will have an important impact for protecting baseflows during the dry season that are significantly beneficial for aquatic and riparian ecosystems. Furthermore, this clause will prevent the disconnection of surface water and groundwater resources due to groundwater overdraft. Scientifically, SGMA is a very complex scientific and methodological challenge that demands adequate climate, water, economic, and operational data of water resources systems, as well as the integration of models. Surface water and groundwater models are coupled with operations, plant physiology, and economic models to estimate the impacts of different strategies in groundwater basins. Economically, SGMA is incentivizing creative thinking to address any undesirable results that can or are already occurring. In some places this legislation will limit the groundwater extraction and, thus, the economic development of certain activities. Other economic activities are likely to emerge as a substitution of agriculture, such as solar energy harvesting.

Scientist, engineers, authorities, water users, and practitioners are thinking out of the box to avoid or mitigate any undesirable result. SGMA is making us think system-wise for basin water management considering hydrologic, social, economic, and environmental aspects of the basin. For a long time, we have thought about operating these aspects together but never with the detail and integration that we are doing it now. For instance, intentional groundwater recharge through agricultural land referred as agricultural groundwater banking (Ag-GB) is a strategy where water released from a reservoir for flood control purposes are diverted into canals and spread out into agricultural fields that have the adequate type soil and crop for short periods of time to intentionally recharge water into the underlying aquifer. Ag-GB is linking agricultural practices, landscape and soil characteristics, plant physiology for water tolerance, and flood management for improving system's storage for future water supply. Recent studies suggest that there is sufficient unmanaged water available to mitigate groundwater overdraft impact in places where groundwater overdraft is happening (Kocis and Dahlke 2017). Furthermore, borrowing a concept from the energy sector, net water metering has been implemented in some basins, giving farmers economic credits of the water recharged in their property from Ag-GB toward their overall groundwater extraction bill from wells. Moreover, in some places of California, Ag-GB and net metering are implemented in conjunction with the use of recycled water for agricultural water supply, deficit irrigation for certain crops, and land fallowing to reduce the overall water demand in groundwater basin. In other regions of the state, reservoirs are operated considering short-term weather forecast (FIRO – Forecast Informed Reservoir Operations), so surface water storage can be maximized to meet human and environmental water needs and decrease the pressure on groundwater resources. Newly formed GSAs are reaching their constituents to get ideas and support implementation of some of these strategies. As you can see, this is the next level of integrated water resources management where every water sources is accounted and managed conjunctively; where water supply, demand, and storage policies are intertwined with economic incentives and regulations; where the community is working side by side with the newly formed regulatory agencies to get the most out of the water scarce resources; and where economic incentives and out of the box strategies are proposed, tested, and implemented.

References

Bork KS, Krovoza JF, Katz JV, Moyle PB (2011) The rebirth of California fish & game code section 5937: water for fish. UCDL Rev 45:809
Clean Water Act (2008) Clean Water Act. EPA's Office
Congress US (1973) Endangered Species Act. US Code 16:1534–1544
Corcoran K (2018) California's economy is now the 5th-biggest in the world, and has overtaken the United Kingdom. https://www.businessinsider.com/california-economy-ranks-5th-in-the-world-beating-the-uk-2018-5. Accessed Sept 2018
DWR (2018) State water project. Department of Water Resources. https://water.ca.gov/Programs/State-Water-Project. 2018
Forbes (2017) California gross state product. https://www.forbes.com/places/ca/. Accessed Sept 2018
GWP (2000) Integrated water resources management, vol 4. Global Water Partnership
Hanak E et al (2011) Managing California's water: from conflict to reconciliation. Public Policy Institute of California, San Francisco
Kadir T, Huang G (2015) Unimpaired flows vs. natural flows to the Sacramento–San Joaquin Delta: what's the difference? In: California water and environmental modeling forum, Folsom, California. Retrieved from http://www.cwemf.org/AMPresentations/2015/Kadir_NaturalFlow.pdf
Kocis TN, Dahlke HE (2017) Availability of high-magnitude streamflow for groundwater banking in the Central Valley, California. Environ Res Lett 12:084009
Littleworth AL, Garner EL (2007) Chapter 3: Surface water rights. In: California water II. Solano Press, Point Arena, pp 37–67
Loucks DP, Van Beek E, Stedinger JR, Dijkman JPM, Villars MT (2005) Water resources systems planning and management; an introduction to methods, models and applications. United Nations Educational, Scientific and Cultural Organization, Delft
Olshansky RB (1996) The California Environmental Quality Act and local planning. J Am Plan Assoc 62:313–330
PPIC (2016) Water for the environment. Public Policy Institute of California, Water Policy Center
Sax JL (1970) The public trust doctrine in natural resource law: effective judicial intervention. Mich Law Rev 68:471–566
Tarlock AD, Tippy R (1969) Wild and Scenic Rivers Act of 1968. Cornell L Rev 55:707
Trelease FJ (1957) The concept of reasonable beneficial use in the law of surface streams. Wyo LJ 12:1
USBR (2018) Central valley project. US. Bureau of Reclamation
Wurbs RA (2006) Methods for developing naturalized monthly flows at gaged and ungaged sites. J Hydrol Eng 11:55–64
Zektser S, Loáiciga HA, Wolf JT (2005) Environmental impacts of groundwater overdraft: selected case studies in the southwestern United States. Environ Geol 47:396–404. https://doi.org/10.1007/s00254-004-1164-3

Chapter 5
International Comparative Analysis of Regulations for Water Markets and Water Banks

María E. Milanés Murcia

Abstract Water markets and water banks are mechanisms to transfer water among different users. A legal and institutional framework is the tool that can protect third parties and ensure the success of the water market and the water bank. A comparative analysis of water markets and water banking in Chile, Australia, and the United States shows how strong regulations have a positive effect on the management of water markets and water banking. The case in Chile reveals that the 1981 Water Code lacks legal protection in areas such as environment, sociology, and the integration management of water resources. This has been a cause of the failure of the water market in Chile. Differently, in Australia, the environment is recognized as a legitimate water user for which states could specifically establish environmental water allocations. The United States has several examples of successful water banks such as the Kansas Water Bank, which has a very exhaustive regulation and has effectively promote water conservation and improve the use of groundwater resources.

Keywords Transfer water · Water banking · International regulations of allocating water

5.1 Introduction

Water transfer mechanisms and regulations to allocate water are essential to ensure the optimal use of water. There are many mechanisms to transfer water, such as water markets, water banking, and legal mechanisms. Among these legal mechanisms to transfer water, forfeiture and abandonment have a large impact on the management of water (Szeptycki et al. 2015). Water markets and banking require a comprehensive legal and institutional framework to ensure water for all users but

M. E. Milanés Murcia (✉)
Sacramento, CA, USA

E. O. Vieira et al. (eds.), *Integrated Water Resource Management*,
https://doi.org/10.1007/978-3-030-16565-9_5

especially to guarantee water for environmental purposes. The main goal of this chapter is to demonstrate how water banks and water markets under regulation can be successful tools to allocate water while protecting the environment.

In general, legal and institutional framework influence the economy in each social and cultural context. Law and economics are instruments that working on the same direction can successfully achieve the policy goals proposed and implement optimal water markets and water banks. Appropriate legal, regulatory, and administrative framework is essential in the management of water resources (Hansen 2015). In fact, "markets are based on a system of enforceable private property rights. Private water markets require secure and transferable property rights, including the right to exclude other users" (FAO 1993).

Water transfer regulation varies from one country to another and specifically in federal countries such as the United States where the laws and regulation are different from one state to another. This situation makes it difficult to develop homogeneous water banking for each basin with water rights able to be transferred from one jurisdiction to another (Hansen 2015). A comparative analysis of different water transfer mechanisms and regulations provides the framework to develop modern water rights based on uses and able to transfer water in conjunctive use.

In Australia, the legal framework developed under the Water Act 2007 and the National Water Initiative (NWI) provides the basic elements to develop an efficient water allocation mechanism that balances water use between competing uses. Water Management Plans are key to establishing strategies, implementing regulations, and allowing flexibility in changing conditions. Under this legal framework, water markets are the basic tool to achieve security to water users in the case of water scarcity, as well as to move water from consumptive to environmental uses (Pilz 2010).

The Water Code 1981 in Chile provided a set of regulations and requirements that allow the private sector to establish their own water market. The Code lacks of legal protection in areas such as environment, sociology, and the integration management of water resources. Within the country some water markets have been successful, while others have resulted in conflicts among users (Navarro-Caballero 2006).

The United States has several examples of successful water bank such as the Kansas Water Bank, which was established to promote water conservation and improve the use of groundwater resources (Peck 1994). The 1991 Drought Water Bank is the most representative example of a successful water bank, with the state of California being the predominant broker for water trades. One of the main goals of this water bank was the protection of fish and wildlife and their habitat, where the efforts of State agencies reduce the impact of drought in the wildlife (Dixon et al. 1993).

This chapter provides the analysis of water transfer regulations and the role of a water bank to allocate water ensuring optimization. It contains (1) an analysis of water markets and water banking as mechanisms to transfer water; (2) an international comparative analysis of water transfer mechanisms in Australia, Chile, and the United States; and (3) results from the international comparative analysis and conclusions.

5.2 Water Markets and Water Banks

The most appropriate type of mechanism to transfer water will be one established by local circumstances (Le Quesne et al. 2007). Sometime, a mix of allocation mechanisms is needed as the solution to allocate water. For example, water banks can be combined with spot trading and the purchase of rights to facilitate strategic reallocation between sectors, such as environmental needs and growing urban demands (Le Quesne et al. 2007). Water markets are only one of the different mechanisms to reallocate water. This section presents an overview of the most relevant mechanisms for water allocation.

5.2.1 Water Markets

According to Colby Saliba and Bush, a water "market" consists of the "interactions of actual and potential buyers and sellers of one or more interrelated water commodities" (Saliba and Bush 1987). These authors distinguish market transfers from other transfer processes. A water market may include the following elements:

1. Water is different from land and improvements, and it has a recognized value.
2. Buyers and sellers are willing to reallocate their water rights voluntarily because it is their best interests.
3. Price and other terms of transfer are negotiable among buyers and sellers. The price can also be negotiated through an institution able to protect buyers, sellers, and third parties' interests (Saliba and Bush 1987).

Transactions can include sale or lease of fee titles, water use permits, conservancy district shares, and project contract rights or conditional water leases for drought year use and exchanges of water rights where a variety of dates and arrangement are used to save water. The waters included in the market transactions are groundwater, native and imported surface water, artificially recharged and recovered water, effluent, and conserved water. The combination of all different sources of water creates connectivity among them as a unitary whole that provides security to develop a unique type of water right able to allocate surface and groundwater in conjunctive use. The key element for a market transaction is that it fulfills the condition of obtaining net benefits while transferring water to a new use (Saliba and Bush 1987).

Water markets can be divided into four categories:

1. Sales involve the transfer of title with benefits, obligations, costs, and risks associated therewith. A sale means the permanent transfer of all legal claims under the right.
2. Leases are a temporary transfer that may be from a single season to many decades. The title remains with the owner; in consequence, the lessee will not incur all benefits, costs, risks, and obligations. It may be a way to lease a surplus amount for a specific use to users who have either a short-term demand for additional

water or a long-term but highly variable demand. In addition, a user's demand for water may be both long-term and stable, but supplies may be highly variable. Leasing provides flexibility according to different circumstances.

3. Options are contracts between buyers and sellers that specify the quantity, price, and other conditions under which water may be transferred. An option may be either a sale or a lease transaction. Options are characterized by flexibility and low cost, and they may be a good alternative when buyers are uncertain as to when additional water will be needed because they provide the certainty of knowing that a determined quantity of water can be bought at a given price.
4. Negotiated adjustments do not cause a direct transfer of water rights. They are agreements under which one or both parties agree to take certain actions that result in the buyer benefiting from increased access to water. They are particularly useful when transfers of water are difficult or expensive, but adjustments may release supplies for other uses, as when junior appropriators of water rights in a stream agree to share water with senior appropriators (Saliba and Bush 1987).

There is an extensive variety of water markets. This subsection presents three of the most common water markets:

1. Open water markets. An open water market is defined as a free market where water rights can be traded for the most part without administrative control and interference. Open water markets can be easily introduced where water rights are privately held. This means that such property rights can be traded (Le Quesne et al. 2007).
2. Spot markets. This type of market is a temporary exchange of water and is very common. Typically, the holder of the water right maintains title to the water and leases it, thus allowing flexibility during drought times. However, this mechanism is short-term and does not guarantee water for the future, as the security of supply is unsure (Le Quesne et al. 2007).
3. Informal water markets. This type of market exists even if not sanctioned by official national policy or law (Le Quesne et al. 2007). These markets have no specific structure to allocate water.

5.2.2 *Water Banking*

In general, a water bank can be defined as a brokerage institution whose purpose is to pool water from one user for rental or sale to other users. A water bank buys and sells water under some set of rules regarding prices and quantities, with the bank acting as intermediary in the water market between buyers and sellers (Le Quesne et al. 2007). This creates an incentive to lower transaction costs and encourages market activity. Moreover, a water bank can regulate social and environmental impact. "[W]ater banking is attractive as it can allow for water to be set aside to ensure ecological flows as part of the trading process" (Le Quesne et al. 2007).

A water bank should be strong enough to ensure that public trust is maintained. It needs defined and secure water rights as well as strong management institutions that can monitor water use and enforce the water rights system (Le Quesne et al. 2007). Water bank institutions in each country and state vary according to the physical and socioeconomic circumstance of each region (O'Brien 1999). An example of a water bank is the Conservancy Bank in the Middle Rio Grande that focuses on preventing permanent severance of water for agricultural uses while at the same time providing water for municipal needs and environmental and in-stream uses (O'Brien 1999). The Middle Rio Grande Conservancy District Revised Rule No. 23 states:

> It is the purpose of the Water Bank to support beneficial use and distribution of water for agriculture and related purposes within the Boundaries of the Conservancy District, to promote the welfare of the Conservancy District and of all the inhabitants and constituents thereof, pursuant to NMSA 1978, §§ 73-14-1 through 73-18-43. This welfare comprises the encouragement of agriculture and is conducive to the public welfare and the conservation of water within the state, including groundwater recharge, maintaining delivery of water to rights holders, and promoting food security, while secondarily providing incidental recreational uses and environmental benefits.

5.3 International Comparative Analysis

This section provides an analysis of some relevant examples of water markets and banking. The countries selected for this analysis are Australia, Chile, and the United States.

5.3.1 Australia

5.3.1.1 Water Markets and Banking in Australia

Australia is located in the driest inhabited continent in the world. This country has designed water management reforms at the local, state, and federal levels. The states have had responsibility for water management since the federation in the 1890s when the people of six British colonies, which are now the states, agreed to unite under one constitution as the Commonwealth of Australia (Pilz 2010) (Australian Government). In the early 1990s, the Commonwealth established changes in water governance: (1) through intergovernmental agreements addressing reform principles and (2) multi-jurisdictional agreements between states governing the shared water resources of the Murray-Darling Basin (Pilz 2010).

The Council of Australian Governments (COAG) on a Water Reform Framework had the goal of implementing a strategic framework to achieve an efficient and sustainable water industry. The first step was the development of water markets and water trading to maximize the economic and social benefits of water. An important step forward was the recognition of the environment as a legitimate water user for which states could specifically establish environmental water allocations. Another achievement was the separation of water regulation and policy authority away from water service providers and the inclusion of water pricing reflecting the full cost including externalities (Dinar et al. 1997; Pilz 2010).

The River Murray Water Agreement of 1914 between the Commonwealth, New South Wales, South Australia, and Victoria dealt with water sharing to help ensure security of supply. It also established an arrangement for sharing costs associated with maintenance and the building of infrastructure such as storage (Kildea and Williams 2010). New South Wales and Queensland also formed the Border Rivers Agreement (1946), which managed the Severn, Dumaresq, Macintyre, and Barwon rivers. In addition and similar to previous agreements, the 1985 Groundwater Border Agreement between South Australia and Victoria was developed. Later, the Murray-Darling Basin Agreement 1987 established the cooperative and institutional basis for managing the quantity and quality of water resources (Kildea and Williams 2010).

In addition to water agreements, Australia has implemented water planning, which is a powerful vehicle to promote changes in water use. The National Water Initiative (NWI) and the Water Act 2007 are the main water planning reforms. The NWI provides guidelines for the state and is designed to complete the modernization of Australia's water management system through markets, regulations, and water plans with ambitious environmental and economic goals. The Water Act 2007 created the Murray-Darling Authority and established the Basin Plan addressing water management in the Murray-Darling Basin. The main success of this plan is that it includes environmental sustainability in the Basin. It focuses on (1) environmental assets, i.e., Ramsar-listed wetlands; (2) key ecosystem functions, i.e., floodplain inundation; (3) the productive base, i.e., salinity reduction in order to keep land productive; and (4) environmental outcomes, i.e., recovery of a specific riparian site (Pilz 2010).

The Basin states and the Commonwealth agreed to the Water Amendment Act 2008 as an addition to the Water Act. The states agreed to give away constitutional powers, especially those needed to carry out water resource planning for the Murray-Darling Basin, to the Commonwealth via the authority. Like the NWI, the main purpose of this amendment was to use planning and markets to meet environmental and economic goals. The Water Act calls for the establishment of "environmentally sustainable limits" (Pilz 2010) on water withdrawal called sustainable diversion limits, which implement a broad rebalancing of water between consumptive and environmental uses (Pilz 2010).

It is important to note that the Water Act and the NWI established the development of water allocation planning and water markets. Unlike other countries, Australia has provided a specific legal framework for water markets. The way NWI allocates water is key to the success of its water market. Management is through (1) dedicating a "pool" of the water resource for consumptive use; (2) dividing that pool into shares and creating permanent water rights based on those shares; and finally (3) determining each year how much water to allocate to each share based on how much water is available that year. The NWI defines permanent water rights as entitlements that are a "perpetual or open-ended share of consumptive pool of a specified water resource" (NWI 2004).

Water planning can be a powerful instrument to promote changes in water usage, especially because of the yearly allocation decisions based on prior water usage. Allocation is defined as "the specific volume of water allocated to water access entitlements in a given season, defined according to rules established in the relevant water plan" (NWI 2004). Water is allocated according to the needs and availability for each year, thus allowing the implementation of reasonable and equitable use of water yearly (Milanés-Murcia 2017).

A plan may determine which pool is filled first with the available water. In general the first pool to fill is river operation water, which may include environmental water; without this pool, the entire system might not be operable. After this, consumptive pool and another environmental pool must be filled. As an example, during the drought of 2008–2009, some holders in the state of Victoria received zero percent allocation based on their usage (Pilz 2010). The composition of water rights in Australia is made up of several elements: a permanent entitlement; a yearly allocation; and some form of use approval specific to a piece of land, where water is kept totally separate from land title (Pilz 2010). The main goal is to achieve a progressive removal of barriers to trade in water. At the regional level, Australia has also developed "Water Allocation Plans" (WAPs) by regions called Natural Resource Management boards. WAPs "set out principles associated with the determination of water access entitlements and for the taking and use of water [in order to achieve] an equitable balance between environmental, social and economic needs [in a sustainable manner]" (NWI 2004). From the local to the basin level, regulation is consistent with the setting of "sustainable limits of water that may be taken from the Basin" imposing an "environmentally sustainable level of take" on water uses throughout the Murray-Darling Basin (NWI 2004).

In addition, water management in Australia includes metering, which makes the yearly volumetric allocation system enforceable. The interconnection between water regulators, policy makers, and water suppliers has led to the reality that most water users in Australia pay to use water (Pilz 2010).

The legal framework developed under the Water Act and the NWI provides the basic elements to develop an efficient water allocation mechanism that balances water use between competing uses. Plans are key to establishing strategies, implementing regulations, and allowing flexibility in changing conditions. Under this legal framework, water markets are the basic tool to achieve security to water users in the case of water scarcity, as well as to move water from consumptive to environmental uses (Pilz 2010).

Australian water rights allow both trades in allocations and permanent transfers. The first type of temporary trading is the trading of volumes within 1 year. The second is trades of water entitlements. Each of these two categories can be divided into high versus low reliability entitlement and allocation trades. Institutional water markets have shown that irrigators are willing to reduce an uncertain supply, using the water markets to ensure enough water for their crops in the short term. Water markets in Australia have enabled farmers to stay on their farms and within their farming community by trading their water rights and giving up some of their irrigation (Pilz 2010).

The Australian case proves the need for an institutional and legal framework strong enough to support water transactions among different users. The water markets have been designed according to legal regulations. This demonstrates the relevance of an optimal legal framework in order to have a successful water transfer system able to guarantee water for different uses, especially the reallocation of water for environmental uses (Milanés-Murcia 2017).

5.3.2 Chile

5.3.2.1 Water Market and Banking in Chile

The Water Code of 1981 provides the legal basis for developing a water market. The code regulates separately land and water to permit the purchase and sale of water. The Water Code defines water as a "national public good" and as "market assets" (Chile Water Code 1981). This allowed for the privatization of water rights, which was done through the granting of rights without cost and in perpetuity. The state does not have power to intervene when a reallocation has been made. Water rights are allocated through the buying and selling of water rights (Ariño and Sastre 1999). The water transfer mechanism is very flexible and allows modification of water uses from the original concession. This means that users do not have to justify any future use, and therefore a water transfer can change the type of use (Andrade Geywitz 1991; Dinar et al. 1997). At the legal level, there is no provision establishing a preference of one type of use over another. This may create conflict when there are several requests for the same use or sector, but it is a free market where supply and demand establish the rules of the game. Therefore, there is a risk that one private agent or sector may gain a monopoly of all water rights (Navarro-Caballero 2006).

Under Article 22 and 141 Chile Water Code, the original concession of water rights is free and without any regulation. The only requirement is the existence of the availability of water resources and that there will be no harm to other users. Water right owners do not pay any tax or cannon in order to acquire or maintain their right. The Chile Water Code thus created a laissez-faire system that established neither a direct legal framework nor a water rights market. It only provides a set of regulations and requirements that allow the private sector to establish their own water market (Navarro-Caballero 2006).

Within the country some water markets have been successful, while others have resulted in conflicts among users. The Limali watershed is an example of efficient water right transfers. It is located in the center-north of the country and contains an important agricultural water market. The main reasons for its success are (1) large water storage capacity, with three dams and canals connected to them; (2) the existence of a strong water users association with an infrastructure able to allocate water; and (3) a climatology rich enough to produce high-quality production. This has contributed to the reallocation of water from inefficient uses to others with higher economic value (Bauer and Orrego 2004).

On the other hand, water markets in the Maule and Bio watersheds are examples where conflict has resulted in inefficient water markets. The main reason for these unsuccessful markets is the coexistence of consumptive and nonconsumptive rights that do not allow for integration and cooperation. This is due to a lack of a clear regulation governing consumptive and nonconsumptive rights in the Water Code. The only applicable legislation for this issue are Articles 14 and 15 of the Water Code (Chile Water Code 1981). These Articles establish that the use of water may not cause damage to other users of the same source and that quantity, quality, and use may not be jeopardized by other users. However, the Chile Water Code does not establish any priority of uses, and this lack of specific regulation has created conflicts and differing political views about how to address both rights (Navarro-Caballero 2006).

In summary, the main issue is the legal and institutional framework, which has not been able to correct market failures. A legal framework would have created clear water rights identifying consuetudinary uses and having an accurate register of them. Clear and specific water rights would have provided security and avoided market failures. The most representative benefits of this system are also the causes of market failure. On one side, water rights as property rights provide security; on the other side, water as a commodity allows the transfer of water from one use to another with higher value.

5.3.3 *The United States*

Water transfer in the United States has been especially developed in the Western States, where several projects have taken place. The US Department of the Interior's Bureau of Reclamation is known for canals, dams, and power plants it has built in 17 Western states (USBR 2018). Water transfer has been used for environmental protection in the Western United States. Streamflow markets have been driven by efforts to improve water quality and to restore flows for endangered fish species. For example, the Bureau of Reclamation began a campaign to lease water for endangered salmon species in the Columbia River Basin (Landry 1998). Another example is the Oregon Water Trust, which established the Water Rights Acquisition Program. Its goal is to "increase stream flows for fish conservation, water quality improvements, or recreational use by purchasing, leasing or in other ways acquiring water rights (permits to take water from a stream) from voluntary sellers, mainly farmers, ranchers and private landowners" (The Oregon Water Trust 1993). Although the number of example in water transfer mechanisms is very large in the United States, this section focuses on the Kansas Water Bank and California water trade systems.

5.3.3.1 Kansas Water Bank

The Kansas Water Resources Appropriation Act 1945 (Kansas Water Appropriation Act 1945) provided the legal framework to develop a water bank. The main purpose of the Kansas Water Bank has been to promote water conservation and improve the use of groundwater resources. This water bank defines the "conservation element as the portion of a deposit that is taken out of use for the duration of the deposit and is not allowed to be withdrawn and used by subsequent users" (Kansas Water Appropriation Act 1945). This provides a minimum of water so as to achieve a sustainable level that guarantees the use of water for the future. The law also defines and establishes safe deposit accounts, which "refers to the deposit of a water right, or portion of a water right, in a water bank for the purpose of having the bank lease water from such water right, or portion of a water right, to another person or entity" (Kansas Water Appropriation Act 1945). The water bank was developed as a mechanism to "redistribute water to areas of growing need or to supplement streamflow and encourage conservation through a market technique" (Sophocleous 2012). This water bank has complex rules for transactions, and the Committee has recommended simplification of regulation of the bank in order to make it available to all users. Although complexity has been a negative factor, these rules represent a comprehensive instrument for development of a legal framework capable of providing sustainability, conservation practices, and improvement of the environment (Milanés-Murcia 2017).

5.3.3.2 California

The Central Valley Project (CVP) is one of the major water conservation developments in the United States. It runs from the Cascade Range in the north to the semi-arid but fertile plains along the Kern River in the south. The original purpose of this project was to protect the Central Valley from water shortages and floods. The CVP contributed to enhancing Sacramento River navigation and domestic and industrial water supplies; it generates electric power, improves recreation facilities, and provides conservation for fish and wildlife, as well as assuring better water quality. The CVP's facilities are essential to provide water to farms, homes, and industry in California's Central Valley (USBR 2018).

The California State Water Project is composed of a water storage and delivery system of reservoirs, aqueducts, power plants, and pumping plants. The goal of this infrastructure system is to store water and distribute it to agricultural water suppliers in Northern California, as well as 29 urban areas, the San Francisco Bay Area, the San Joaquin Valley, the Central Coast, and Southern California. The use of this water is dedicated 70 percent to urban users and 30 percent to agriculture. The California Department of Water Resources manages this project. The project improves water quality in the Delta, controls floods, provides recreation, and enhances fish and wildlife (DWR 2018).

> The CVP reaches some 400 miles, from the Cascade Mountains near Redding in the north to the Tehachapi Mountains near Bakersfield in the south. Consists of 20 dams and reservoirs, 11 powerplants, and 500 miles of major canals, as well as conduits, tunnels, and related facilities. Manages some 9 million acre-feet of water. Annually delivers about 7 million acre-feet of water for agricultural, urban, and wildlife use. Provides about 5 million acre-feet for farms -- enough to irrigate about 3 million acres, or approximately one-third of the agricultural land in California. Furnishes about 600,000 acre-feet for municipal and industrial use--enough to supply close to 1 million households with their water needs each year. Dedicates 800,000 acre-feet per year to fish and wildlife and their habitat and 410,00

acre-feet to State and Federal wildlife refuges and wetlands, pursuant to the Central Valley Project Improvement Act (CVPIA) (USBR 2018).

The CVP provides water for six of the top ten agricultural counties in the nation's leading farm state. "It has been estimated that the value of crops and related service industries has returned 100 times Congress's $3 billion investment in the CVP. About 60 percent of the cost of the CVP was allocated to irrigation and municipal and industrial water with the remainder to other beneficial uses" (DWR 2018). This infrastructure allows the development of water allocation among different users and therefore establishes the conditions for successful water markets and banking (Milanés-Murcia 2017).

The California State Water Code and current Federal law in Section 3405(a)(1) of the Central Valley Project Improvement Act of 1992 (CVPIA) (Public Law 102-575, 106 Stat. 4709) establish the conditions under which CVP water is transferable.

The "Bureau of Reclamation is working with the California Department of Water Resources in developing consistent evaluation criteria for a long-term, programmatic water transfer program designed to provide for water transfers from State and Federal contractors North of the Delta to contractors South of the Delta. These transfers will continue to be subject to the consumptive and beneficial use requirements in the State Water Code" (Conner 2009).

The Bureau proposed a new legislation, which intends to improve flexibility and efficiency for water management in the Central Valley. The Commissioner of the US Department of the Interior stated before the Energy and Natural Resources Committee Subcommittee on Water and Power US Senate regarding S. 1759 Water Transfer Facilitation Act of 2009 that:

When they are done right, water transfers move water from willing sellers to willing buyers in transactions that can improve economic well-being, increase efficiency in water use, and protect against negative externalities. There are many situations where water transfers during periods of drought can be used to ensure that available water is used in areas where it is most needed, and S. 1759 is aimed at facilitating these efficient water transfers. We recognize the potential of voluntary water transfer as a mechanism to increase flexibility into our water management system and respond to changes in available water resources. However, we are also committed to implementing review processes for all water transfers that will effectively protect the broad range of interests that can be impacted by changes in water use. Our goals as a Department include ensuring efficient use of available water infrastructure as well as maintaining vibrant communities and protecting the environment (Conner 2009).

The 1991 Drought Water Bank is the most representative example of a successful water bank, with the state of California being the predominant broker for water trades. One of the main goals of this water bank is the protection of fish and wildlife and their habitat. DWR and the Department of Fish and Game work to reduce the impact of drought in the wildlife. These efforts resulted in modifications to SWP operation and Water Bank transfers. As an example, the Bank kept water in the Shasta Reservoir for temperature control for the 1991 fall and winter salmon run (Dixon et al. 1993).

Assembly Bill (AB) 9 established an authority to enter into contracts with DWR or other water suppliers for the transfer of water outside their service area. In addition, AB 10 provided that no temporary transfer of water for drought during the period 1991 and 1992 would affect existing water rights (Dixon et al. 1993). Farmers agreed to fallow their land and credit the conserved water to the Bank. Contracts involving the substitution of groundwater for surface water, direct groundwater pumping, and reservoir storage withdrawals were negotiated (Dixon et al. 1993). Only a few contracts addressed pumping of groundwater, and less than 10,000 acre-feet were transferred to the Water Bank. This situation raised the question of whether the water was "new" non-surface water. Water was "new" if it had been made available to the state as part of the Water Bank. In order to ensure that the Bank received new water, well logs for each well were reviewed (Dixon et al. 1993). Similarly, fallowed lands were monitored to confirm compliance with Drought Water Bank contracts.

It was difficult to determine a fair price for water, and in general, the price of water did not reflect the value for consumers. However, on this occasion the value of water was an important aspect in establishing the price. The goal was to offer "a price that would yield a net income to the farmer similar to what the farmer would have earned from farming plus and additional amount to encourage the farmer to enter into a contract with a new and untried Water Bank" (Dixon et al. 1993).

The flexibility of the SWP and CVP permitted a successful implementation of this water bank, which reduced negative third-party impacts, broke down institutional barriers to water transfers, established a consistent water price meeting critical needs, and minimized risk (Dixon et al. 1993).

The Kern Water Bank is another example of water banking. The Monterey Agreement in 1994 between the California Department of Water Resources and representatives of the agricultural and urban contractors was the basis for this groundwater bank. The principles of this agreement established the state's transfer of ownership of the Kern Fan Element (KFE) property to Kern County Water Agency (KCWA) and then to the Kern Water Bank Authority (KWBA), for the use of local agencies and groundwater banks. Among the main benefits of this water bank were the return of waterfowl and an increase in wetland habitat. More than 40 new species of birds appeared in the area after this water banking was established (Deltarevision 2003).

Another example of a water bank is the 2009 Drought Water Bank. Water rights holders from upstream in the Sacramento-San Joaquin Delta sold water to DWR. The State Water Project and Central Valley Project were used to transfer water, thus providing water to suppliers that were at risk of experiencing water shortages in 2009 due to drought conditions and fulfilling required supplemental water needs to meet anticipated demands (DWR 2010).

5.4 Results from the Comparative Analysis

From the analysis, the Australian case proves the need for an institutional and legal framework strong enough to support water transactions among different users. The water markets have been designed according to legal regulations. This demonstrates the relevance of an efficient legal framework in order to have a successful water transfer system able to guarantee water for different uses, especially the reallocation of water for environmental uses. The Australian system gives priority to the law over the economic aspect, thus establishing a flexible and strong legal and institutional framework able to allocate water and protect the environment during drought times.

The failure of the Chilean water market is a lack of legal regulation that allowed the economic aspect to take preference over the legal. This left the market without legal protection in areas such as environment, sociology, and the integration management of water resources.

The Kansas Water Bank is an example of a water bank in the United States and shows how legislation can establish conservation practices and increase flexibility to water users through groundwater management. A successful water transfer's system must have an appropriate legal and institutional framework able to regulate and guarantee all contractual processes and transaction costs.

In California the State agencies have been the key element to ensure the successful implementation of water banking, such as the 1991 Water Bank, which was able to protect the environment during a drought. When the institutional framework is trusted by the users, the implementation of the regulation will be effective like in the case of California.

5.5 Conclusions

Water markets and specially water banks have been demonstrated to be optimal to encourage transfers and promote conservation. The success of these tools is subjected to the establishment of a capable and flexible legal framework where an institution has authority to regulate legal framework based on social, physical, and economic circumstances as well as setting a fixed price that avoids uncertainty and risk. In addition, water banks are more politically accepted than private markets. In countries where water banks are established at the local level, there is a potential willingness to agree in the implementation of international water banking agreement in the management of international basins between the riparian countries while protecting the environment, specially during drought periods.

References

Andrade Geywitz C (1991) Reforma de la Constitución Política de la República de Chile de 1980

Ariño G, Sastre M (1999) Leyes de aguas y política hidráulica en España Comares, Granada

Bauer CJ, Orrego JP (2004) Canto de sirenas: el derecho de aguas chileno como modelo para reformas internacionales. Bakeaz

Chile Water Code (1981) Decreto con Fuerza de Ley N. 1.122. Santiago, Chile

Conner M (2009) Statement of Michael L. Connor. US. Bureo of Reclamation

Dinar A, Rosegrant MW, Meinzen-Dick R (1997) Water allocation mechanisms: principles and examples. The World Bank, Washington, D.C.

Dixon LS, Moore NY, Schechter SW (1993) California's 1991 drought water bank. Rand, Santa Monica

Deltarevision E (2003) Study of transfer, development, and operation of the Kern Water Bank. Available at http://www.deltarevision.com/2010%20docs/kern_water_bank_study.pdf

DWR (2010) California's drought of 2007–2009 an overview. Department of Water Resources. Available at https://water.ca.gov/-/media/DWR-Website/Web-Pages/Water-Basics/Drought/Files/Publications-And-Reports/Californias-Drought-of-200709An-Overview.pdf

DWR (2018) State water project. Department of Water Resources. https://water.ca.gov/Programs/State-Water-Project. 2018

FAO (1993) The state of food and agriculture, vol 26. Food & Agriculture Organization, Rome. Available at http://www.fao.org/3/t0800e/t0800e.pdf

Hansen K (2015) Water markets: from theory to practice (with focus on the USA). In: Dinar J, Schwabe K (eds) Handbook of water economics. Edward Elgar, Cheltenham, pp 355–372

Kansas Water Appropriation Act (1945) K.S.A. 82a-701 through 82a-737 and 82a-740 through 82a-742 and K.S.A. 42-303 and 42-313

Kildea P, Williams G (2010) The Constitution and the management of water in Australia's rivers. Sydney L Rev 32:595

Landry C (1998) Market transfers of water for environmental protection in the western United States. Water Policy 1:457–469

Le Quesne T, Pegram G, Heyden CVD (2007) Allocating scarce water: a WWF primer on water allocation, water rights and water markets. In: Allocating scarce water: a WWF primer on water allocation, water rights and water markets. WWF-UK. Available at http://assets.wwf.org.uk/downloads/scarce_water.pdf

Milanés-Murcia M (2017) Proposed international legal and institutional framework for conjunctive management of surface and groundwater along the us–Mexico border region. In: Management of transboundary water resources under scarcity: a multidisciplinary approach. World Scientific, pp 117–157. Availlable at https://doi.org/10.1142/9896

Navarro-Caballero TM (2006) Los instrumentos de gestión del dominio público hidráulico. Universidad de Murcia

NWI (2004) Intergovernmental agreement on a National Water Initiative-between the Commonwealth of Australia and the state governments of New South Wales, Victoria, Queensland, South Australia, the Australian Capital Territory and the Northern Territory. National Water Initiative. Available at https://www.pc.gov.au/inquiries/completed/water-reform/national-water-initiative-agreement-2004.pdf

O'Brien M (1999) Possible solutions: policy tools to achieve flexibility to meet new conditions, preliminary thoughts for coping with future droughts. Nat Resour J 39:175–177

Peck JC (1994) The Kansas water appropriation act: a fifty-year perspective. U Kan L Rev 43:735

Pilz RD (2010) Lessons in water policy innovation from the World's driest inhabited continent: using water allocation plans and water markets to manage water scarcity. U Denv Water L Rev 14:97

Saliba BC, Bush DB (1987) Water markets in theory and practice Market transfers, water values, and public policy (Westview studies in water policy and management, No 12)

Sophocleous M (2012) The evolution of groundwater management paradigms in Kansas and possible new steps towards water sustainability. J Hydrol 414:550–559

Szeptycki LF, Forgie J, Hook E, Lorick K, Womble P (2015) Environmental water rights transfers: a review of state Laws prepared by water in the west for the National Fish and Wildlife Foundation, Washington, D.C.

The Oregon Water Trust (1993). http://walker-foundation.org/net/org/info.aspx?s=24518.0.69.5316

USBR Central Valley Project. US. Bureau of Reclamation (2018)

Chapter 6
Managing Water Differently: Integrated Water Resources Management as a Framework for Adaptation to Climate Change in Mexico

J. Pablo Ortiz-Partida, Samuel Sandoval-Solis, Jesús Arellano-Gonzalez, Josué Medellín-Azuara, and J. Edward Taylor

Abstract Climate change will affect water availability and its management, with more frequent and extended droughts, more severe floods, and lower water quality. Water allocation policies, regulations, and infrastructure in Mexico are not designed for changing future climate conditions. This chapter reviews the implications of climate change for water resources systems in Mexico and evaluates how management strategies from California can serve as potential adaptation schemes toward an Integrated Water Resources Management framework in Mexico.

Keywords Water allocation policies · Climate change · Water management strategies

J. P. Ortiz-Partida (✉)
Hydrologic Sciences Graduate Group, University of California, Davis, Davis, CA, USA
e-mail: joportiz@ucdavis.edu

S. Sandoval-Solis
Department of Land, Air and Water Resources, University of California, Davis, Davis, CA, USA
e-mail: samsandoval@ucdavis.edu

J. Arellano-Gonzalez
Agricultural and Resource Economics, University of California, Davis, Davis, CA, USA
e-mail: gonzalez@primal.ucdavis.edu

J. Medellín-Azuara
School of Engineering, University of California, Merced, Merced, CA, USA
e-mail: jmedellin-azuara@ucmerced.edu

J. E. Taylor
Agricultural and Resource Economics, Social Sciences and Humanities, University of California, Davis, Davis, CA, USA
e-mail: jetaylor@ucdavis.edu

E. O. Vieira et al. (eds.), *Integrated Water Resource Management*,
https://doi.org/10.1007/978-3-030-16565-9_6

6.1 Introduction

In Mexico, reduction in water availability as consequence of climate change not only compromises water reliability for industries and agriculture but also augments the challenge to provide the most basic human right, drinking water. The understanding of the magnitude and extent to which climate change will affect human and natural systems is critical to better design policies that prepare for effective adaptation.

According to the Intergovernmental Panel on Climate Change (IPCC), rising temperatures can reduce renewable surface water and groundwater resources, vital inputs for people, agriculture, industry, energy production, and aquatic and riparian ecosystems (IPCC 2014). As a result, Mexico can experience major impacts on water availability and supply, compromising food, water, energy, and environmental security. In addition, changes in climate patterns are expected to increase drought years, having particular impacts on agriculture, and at the same time intensify rain events, augmenting flood risks in certain areas (Herrera-Pantoja and Hiscock 2015).

Mexico is already facing other water problems that are expected to aggravate as climate changes occur. The Water Advisory Council (CCA 2016), using data from the National Water Commission of Mexico (CONAGUA), noted a series of facts to be considered to improve the current situation and prepare for upcoming water challenges: (1) 22.7% of surface water is heavily contaminated; (2) national potable water and sewage coverage are 91.6% and 90.2%, respectively; (3) 77% of water is used by agriculture; (4) conveyance and distribution of water have an efficiency of 86% and 76%, respectively; (5) 16.2% of aquifers are under overdraft conditions; (6) about 40% of urban water is lost through system leaks; and (7) less than 50% of wastewaters are treated.

Considering the extent of the problem, Integrated Water Resources Management (IWRM) is a powerful and important framework to examine adaptation to climate change. The GWP (2000) defines IWRM as "a process which promotes the coordinated development and management of water, land and related resources in order to maximize economic and social welfare in an equitable manner without compromising the sustainability of vital ecosystems." In the climate change context, an IWRM framework would consider the different systems involved and search for systematic solutions.

6.1.1 *Hydrologic-Administrative Regions as Units for IWRM Implementation*

México has an extensive territory with varied climatic conditions where water availability does not match water demands. The two thirds of the territory with the highest economic development (north, northwest, center Mexico) is also the area that displays the lowest mean annual precipitation (Fig. 6.1) (CONAGUA 2015).

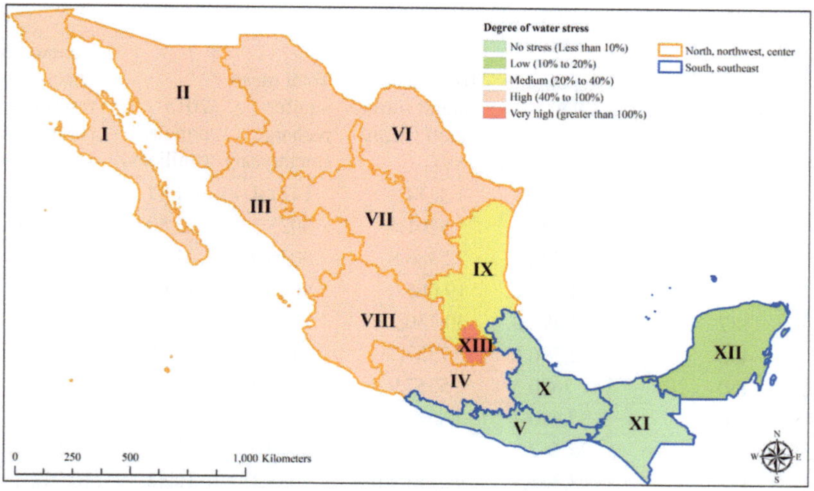

HAR code	Name	HAR code	Name
I	Península de Baja California	VIII	Lerma-Santiago-Pacífico
II	Noroeste	IX	Golfo Norte
III	Pacífico Norte	X	Golfo Centro
IV	Balsas	XI	Frontera Sur
V	Pacífico Sur	XII	Península de Yucatán
VI	Río Bravo	XIII	Aguas del Valle de México
VII	Cuencas Centrales del Norte		

Fig. 6.1 Hydrologic-Administrative Regions and water stress in Mexico. (Adapted from CONAGUA 2015)

The most contrasting examples are the Baja California Peninsula which on average receives 168 mm of precipitation, while in the southeast, in the area of Tabasco and Chiapas, the normal precipitation is around 1842 mm.

Between these extremes, there is a spatial distribution of available water resources in the country, and each particular area requires its own analysis and solutions (CONAGUA 2015).This spatial and temporal distribution of water resources represents different challenges for different areas. With the purpose of water management and preservation of Mexico's water resources, CONAGUA designated 13 Hydrological-Administrative Regions (HARs) (Fig. 6.1).

The HAR with less renewable water resources (XIII) is the second most populated (as it includes Mexico City), consequently having the lowest per capita renewable water in the country (Table 6.1) and catalogued as an area of very high water stress. A situation of water stress arises when the percentage of water diversion is above 10% of the annual renewable water resources. The degree of water stress varies as this ratio increases; above 40% is considered as a high water stress, and above 100% it is catalogued as a very high water stress. Eight of the 13 HARs are at or

Table 6.1 Water availability by HAR

HAR	Normal precipitation from 1971–2000 (mm/year)	Renewable water resources (mcm/year)	Total mean natural surface runoff (mcm/year)	Total mean aquifer recharge (mcm/year)	2015 population (millions)	Renewable water resources (m^3/person/year)
I	169	4681.7	3244.3	1437.4	4.37	1271.2
II	445	8226.7	5201.4	3025.3	2.80	3168.1
III	747	25,422.6	22,386.8	3036.0	4.47	6197.6
IV	963	24,276.1	19,665.0	4611.8	11.69	2251.5
V	1187	32,492.0	30,730.3	1761.7	5.02	7473.3
VI	438	12,796.5	7357.7	5438.8	12.15	1176.7
VII	430	7620.3	5357.9	2262.4	4.52	1829.9
VIII	816	35,680.7	27,524.8	8155.8	23.89	1680.8
IX	914	25,562.8	23,543.4	2019.6	5.23	5109.3
X	1558	98,301.5	94,213.4	4088.1	10.48	10,047.3
XI	1846	157,743.9	138,541.8	19,202.2	7.57	23,519.5
XII	1218	29,338.6	4036.8	25,302.0	4.52	7473.3
XIII	606	3535.8	1432.5	2103.4	23.01	165.3
Nat'l	872	465,679.1	383,235.6	82,443.7	119.71	4285.4

Source: Statistics on water in Mexico reports from their first edition on 2003 to 2015
mcm million cubic meters

above the high water stress classification, 1 on medium, 1 on low, and only 3 are not under stress (Fig. 6.1) (CONAGUA 2015). Under those conditions, excessive groundwater extractions help to meet water demands but generate overdraft on the aquifers (more extracted water than naturally recharged). In Mexico, more than 15% of aquifers are under overdraft conditions (CONAGUA 2015). Groundwater overdraft brings complications, such as seawater intrusion in coastal zones, land subsidence that affects infrastructure, and reduction of water quality and water supply, especially during drought periods.

Groundwater use creates daunting challenges in some of the largest cities of Mexico, particularly Mexico City. The large impermeable extent of the city (asphalt and concrete) makes it hard for the already overdrafted aquifers to recharge. The constant land subsidence (5–10 cm/year) due to excessive groundwater extraction reduces storage capacity and damages sewage and water supply systems (along with other infrastructure). Water infrastructure is in such bad condition that 40% of potable water is estimated to be lost before reaching its destination (De la Peña et al. 2013). The conveyance systems for wastewater are insufficient and inefficient; only about 30% of wastewater from Mexico City is currently treated (60% once the Atotonilco Wastewater Treatment Plant Project is finished) (De la Peña et al. 2013).

Most of the consumptive water use in the country (77%) is for agriculture (CONAGUA 2015). Therefore, efforts to develop adaptation strategies for water supply shocks resulting from climate change need to have a special focus on the possible impacts on agricultural systems.

In the Pacífico Norte HAR (mainly the state of Sinaloa), agriculture is the main economic sector and represents the largest agriculture industry for Mexico. This region produces not only the breadbasket for Mexico but also exports fruits and vegetables to the United States. This region has a high irrigation efficiency; however, this is not a common pattern, and a large share of the irrigation systems in Mexico are still surface irrigation systems (wild flood and furrow). An important amount of water used in these surface irrigation systems does not benefit the crops; instead, it is lost to evaporation or infiltrated into aquifers. Water lost via infiltration can be recovered by pumping; however, more energy is needed to recover this water.

The remaining uses are urban (14%, domestic and municipal) and industrial (9%, including hydropower). Urban water use requires a constant water supply throughout the year. Naturally, there is temporal water availability, resulting in a mismatch of water supply and water demand for this use. Typically, large cities, e.g., Mexico City, Guadalajara, and Monterrey, meet their water demand through water imports from other basins (Cutzamala, Lerma-Chapala, and Cerro Prieto). As a result, sustainable use of water resources is threatened by urban demands requiring imported water from neighbor basins.

Since the beginning of the twentieth century, the model to meet increasing water demands across the country has been mainly focused on infrastructure development: reservoirs, diversion channels, extraction wells, and water delivery systems. As consequence, there has been a continued degradation of ecosystems as the environmental and social impacts of some of these projects altered natural flow patterns intensively and through extensive regions (i.e., reservoirs that flood thousands of hectares). Environmental protection has been focused in specific portions of rivers and mangroves. However, protection throughout rivers is not present or enforced. River fragmentation has occurred and will continue in the form of construction of large reservoirs, canals, and irrigated land. Typically, environmental protection and economic development are seen as opposed activities. Nonetheless, novel techniques have proven the contrary; it is possible to promote economic development while conserving or restoring aquatic and riparian ecosystems; one potential example is the reoperation of Luis L. León Reservoir in the Big Bend Reach of the Rio Grande/Bravo basin (Ortiz-Partida et al. 2016).

Water quality also raises concern related to human health and the conditions of aquatic and riparian ecosystems. In spite of regulation that forbids discharge of raw water into rivers (CONAGUA 2016), unfortunately, this practice still exists. Problems are not only related to the organic content in water (BOD and coliforms) but also to other water quality parameters such as ammoniac nitrogen, nitrites, nitrates, pH, and, in some cases, heavy metals.

Water problems in Mexico are very diverse and thus need to be addressed by considering a variety of adaptation strategies. This research identifies some of the human welfare implications from climate change in Mexico and a series of adaptation strategies that would be applicable to transition from the current situation toward an IWRM.

6.2 Human Welfare Implications

Individual water problems lead to different human welfare implications. Human welfare is compromised when there are negative aspects on the general condition of a population in terms of diet, housing, healthcare, or education. According to CONAGUA (2015), about 92% of the population has access to potable water; however, in Mexico, potable water is not synonymous with drinking water. Under "drinking" water coverage, CONAGUA (2015) considers "all those who have tap water in their household, outside their household, but within their grounds, from a public tap or from another household." However, this definition does not specify that the water is indeed drinkable. In a population of 120 million, about 10 million people do not have access to tap water (drinkable or not), not even from their neighbors. Climate change is expected to increase the number of people without access to drinking water given reduced water supply and the impacts on water quality that facilitate conditions for waterborne diseases. If we consider water quality parameters outside regulation, the number of people without access to drinking water becomes much higher. Robles et al. (2017) sampled 39 houses within 13 municipalities in the State of Mexico and found that 38 had at least 1 parameter outside the permissible limits under current Mexican regulations.

The agricultural sector will also be highly affected by climate change due to an increment in crop water demand, droughts, water scarcity, and changing climate conditions. Given increases in temperature, more water will be needed to meet crop evapotranspiration requirements. This condition will put agriculture at a higher risk because droughts are expected to be more severe and frequent, affecting the water availability for the agriculture. Population growth will reduce or limit water availability for agriculture. In addition, changing climate conditions can bring new diseases to crops and livestock. Farmers and rancher workers are usually in the lower quartiles for annual income in Mexico. Climate change will put at risk this economically disadvantaged communities and the economic viability of corporations and family companies, exacerbating the economic vulnerability of these groups.

Floods are expected to be more frequent and severe. Furthermore, floods will happen in locations where they did not use to occur. Population will be at higher risk of floods, greater likelihood of losing life in places where water reclaims its floodplains, as well as losing family assets such as homes and other material valuables. Large infrastructure will be compromised, demanding more investment or a change in policy such as incentivizing local infrastructure for water detention and recharge.

Poor land management such as deforestation can impact the quantity and quality of water sources putting at risk people and the ecosystems. Without the protection of water sources, such as forest, springs, rivers, lagoons, and aquifers, water quality is expected to decrease. These conditions will be worsened by climate change, due to a higher likelihood of forest fires triggered by severe and frequent droughts affecting forest sustainability. Communities depending on local water sources will be at risk of having a reduction or no water available to meet their needs. In addition, raw water will continue to be discharged in rivers, affecting the ecosystems and the populations that rely on these resources at downstream areas. Water scarcity will be translated into less or no water in rivers for sustaining aquatic and riparian ecosystem, as well as less water for dilution of contaminants. Rivers will be fragmented by infrastructure, such as dams, to harvest water needed to meet human requirements. However, this will come at a high price for environmental degradation and/or extinction of certain species.

6.2.1 Climate Change and Agriculture in Mexico

In Mexico, agriculture represents around 3% of GDP and employs around 13% of the total working population (INEGI 2016a, b). 37.5% of the total population lives in the rural sector in localities with up to 15,000 inhabitants. In Mexico, irrigated agriculture accounts for 77% of the freshwater use. Irrigated agriculture represents only 25.9% of the total agricultural area and generates 56.8% of the total commercial value of agricultural commodities. Yields per hectare on irrigated agriculture are up to 3.3 times higher than those from rainfed agriculture (CONAGUA 2015).

Agriculture in Mexico is largely heterogeneous. On one hand, the vast majority of small-scale agricultural producers (66% of all agricultural producers with fewer than 5 ha) farm staples for own consumption and marginally participate in the market for agricultural commodities. These producers are highly sensitive to climate uncertainty as they mostly rely on rainfall as their primary water source, particularly in the south and southeast regions of Mexico. On the other hand, large-scale producers that form the bulk of irrigated agriculture are located in the drier areas of central and northern Mexico. These farmers have easier access to credit, insurance, and new technologies, and their production decisions respond primarily to domestic and international market demands (only 6% of total agricultural producers). The rest are middle-scale producers transitioning toward higher levels of productivity (Monterroso Rivas et al. 2015).

Agricultural producers are expected to experience the effects of climate change differently. Mendelsohn et al. (2010) estimate that by 2100, agricultural land values in rural Mexico will decrease by roughly 50% under three different climate change scenarios. In all scenarios, the authors find that climate change will be more detrimental to irrigated farms than to rainfed farms. Also, rising temperatures will be more harmful to irrigated farms, while precipitation decreases will be more damaging to rainfed farms. Galindo et al. (2015) show similar findings. Their study reports that an increase of 2.5 °C and a simultaneous reduction in precipitation of 10% cause net revenue average losses ranging from 36% to 55% and 14% to 25% for irrigated and rainfed farms, respectively. Regardless of the farm type, climate change effects are expected to be detrimental for agriculture in Mexico, and thus, an increase in rural poverty levels is expected. Lopez-Feldman (2013), employing two climate change models, estimated that by 2100, rural poverty levels in Mexico might increase from currently 45% up to 54% under the most severe climate change scenario. He also found that poverty impacts will be differentiated by region. In the south-southeast, poverty is expected to reach levels above 70%, while in the northwest, where most of the entrepreneurial agriculture is located, poverty levels are practically unaffected continuing to be around 20%.

Yunez-Naude and Rojas-Castro (2008) provide results on the importance of water provision and availability for agricultural production. Using a general equilibrium approach, the authors estimate that a 50% reduction in water supply would decrease agricultural production by 9.2%. Irrigated agriculture would suffer the most with a decrease of 17.9%, while rainfed agriculture would have a small increase of 2.9%. As expected, regions where agriculture is mainly rainfed would experience the least damages. Virtually the production of every crop cultivated in irrigated areas would decrease with maize and beans suffering the largest decreases, 24.3% and 18.9%, respectively. Crop production in rainfed areas will slightly increase in response to increased crop prices, but the increase would not be enough to offset the losses of irrigated agriculture. As a result, imports of agricultural products might increase.

FAO-SAGARPA (2014) estimate that by 2050, 25 states (out of a total of 32) will suffer some degree of profit loss, but 11 will have losses greater than 50%. By 2099, the number of states with losses higher than 50% increases to 20. This analysis also predicts that over the course of the century, maize and bean production, the two most important staples in Mexico's diet, will tend to decrease, particularly in the southern and northwest states, the higher-producing regions. Grassland will also decrease due to decreases in precipitation thus affecting the production of beef and dairy products. Similarly, the production of wheat and fruits will also decrease.

With lower agricultural incomes and limited adaptation strategies, agricultural households will likely opt out of agriculture. By decreasing agricultural productivity, climate change might create a mass of rural workers seeking to make a living from employment in other sectors. Feng et al. (2010) estimate that by 2080, climate change is estimated to induce 1.4–6.7 million adult Mexicans (or 2% to 10% of the current population aged 15–65) to migrate as a result of declines in agricultural productivity alone. Hunter et al. (2013) showed that in historical sending regions of Mexico's, dry years significantly increases the likelihood of at least one member of the household to migrate to the U.S. by 40%. Multi-year droughts increase this likelihood by 75%. In contrast, wet years significantly decrease the odds of US migration by 35%. Similarly, Jessoe et al. (2018) find that extreme heat shocks increase migration domestically from rural to urban areas by as much as 1.4% and internationally to the United States by as much as 0.25%. Extreme heat may also decrease local wage and off-farm employment by up to 1.4%.

6.2.2 *Climate Change and the Environment in Mexico*

Mexico has a great biodiversity as a country and contains a vast number of ecosystems whose protection is important for the entire world (CONABIO 2016). Water resources management for environmental purposes has not been recognized as a need until recent years, when environmental degradation has been evident in terms of decreased water quality and loss of ecosystems. There have been individual efforts to improve the environmental condition of rivers, lagoons, and estuaries. In the Lerma-Santiago-Pacífico hydrologic region, a comprehensive study was done to determine water allocations for different users while sustaining adequate levels in the Chapala lake to prevent it from completely draining (DOF 2006). These studies ended up in a regulation that stablishes the water allocation for every water user in the basin and water quality restrictions for water discharge into the river. In 2012 a binational agreement (Minute 319) was signed to provide environmental flow pulses for restoring habitat in the Colorado Delta (IBWC 2012). This was an important accomplishment of the Colorado River restoration efforts of both countries. Also in 2012, the federal government developed some guidelines for determining environmental flows throughout the Mexican territory (NMX-AA-159-SCFI-2012 2012). These guidelines are meant to support water resources management at the voluntary basis within each hydrologic region. This is a small first step toward including environmental flows into IWRM.

6.2.3 Climate Change and the Urban Sector in Mexico

Models developed for some areas in Mexico show that despite increasing temperatures and reduced water availability, heavy rains may exceed flooding thresholds, augmenting the risk of lives losses and economic damage (Herrera-Pantoja and Hiscock 2015).

Urban and rural populations, agriculture, and industry are increasing their water use and subsequently augmenting their wastewater discharges. When the wastewater is discharged to a stream or water body without treatment, it compromises the water use for agriculture, fishing, recreation, and drinking. Untreated wastewater discharges are common in Mexico, and it's a consequence of a lack of coordination between water users and authorities (De la Peña et al. 2013). Better wastewater management in terms of recollection, conduction, treatment, and discharge is necessary to stop water resources depletion, riparian and aquatic ecosystem degradation, soil contamination, and an overall impact in food security.

Wastewater treatment is a crucial factor to improve water security as it not only prevents the contamination of streams, water bodies, and soils; it also reduces the instream and groundwater demands from some industries and agriculture by recycling treated water.

6.3 Theoretical Framework and Policy Responses

A bottom-up approach with an IWRM framework is suggested to address the challenges that climate change imposes on water resources in Mexico. A bottom-up approach means that stakeholders provide feedback for water resources planning at the local level and authorities are in charge of putting together the feedback of many stakeholder groups from different local regions into a comprehensive basin-wide plan (Loucks et al. 2005). An IWRM framework is a process recommended as a way to manage all water sources (river, lagoons aquifer, spring, recycled water, etc.) to meet urban, industrial, agriculture, and environmental water needs while maximizing economic and social welfare and the sustainability of ecosystems. This framework must be flexible, adaptable, and responsive to needs at the local, regional, and basin level. Shared vision planning is also recommended as a way to show the needs of other competing users during the planning and execution process. This can help to achieve water security in a sustainable manner.

IWRM covers a portfolio of strategies that incorporates different disciplines to holistically manage water resources for improving water supply reliability while protecting the environmental integrity of the basin (Table 6.2). The state of California in the United States is an example of an extended area where multiple strategies have been investigated and applied. California has implemented actions from a comprehensive water portfolio, and many experiences from California can be used to help Mexico to improve water supply reliability and prevent some of the environmental problems expected to be intensified with future climate change.

Some strategies have proven to be successful, and others need more time before having results or further research for its application. Successful strategies include reservoir reoperations, groundwater banking, use of recycled water, conjunctive use of surface water and groundwater, coordinated water extractions for frost protection, and others.

For instance, for California's Russian River, currently there are studies exploring the feasibility of forecast-informed reservoir operations (FIRO) for enhancing

Table 6.2 Integrated water resources management portfolio

Objectives	Activities/strategies
Reduce water demands	Improve agricultural and urban water use efficiency with a constraint on water right extractions, change to crops with less water demand, reduction in cropping area
Improve operational efficiencies and transfers	System's reoperations, build or modify infrastructure, and stablish water transfers
Increase water supply	Conjunctive management of surface and groundwater, desalination, recycled water, increased groundwater recharge
Improve water quality	Drinking water treatment and distribution, groundwater remediation, pollution prevention, wastewater treatment, urban runoff management
Responsible planning and management of resources (stewardship)	Economic incentives, ecosystem restoration, coordinated land use planning and water resources management, educational and recreational activities
Improve flood management	Design resilient flood protection systems, integrated water supply and flood protection management, forecast-informed reservoir operations
Increase support and integration activities to reduce uncertainty	Regional water planning, improve data and tools, develop research and sciences

reservoir storage during the rainy season while protecting urban settlements from flood events. This type of analysis can be utilized and adapted to reservoir management within the Mexican context. Mexican institutions that can benefit from this knowledge and collaboration include CONAGUA, SEMARNAT, SAGARPA, INECC, IMTA, basin councils, irrigation district, state and municipal water agencies, and NGOs, among others.

Another example is California's Pajaro Valley Water Management Agency, in Pajaro Valley, which implemented a groundwater management plan that considered water conservation, expansion, and new infrastructure, as well as tier water prices. These strategies were proposed, discussed, analyzed, and approved in a decision-making process that followed a bottom-up approach (PVWMA 2013). The decision-making process ended with a planning document that specified funding, implementation, and strategies for reducing seawater intrusion; many of these strategies are now under execution. Specifically, the implementation of the water conservation strategy has reduced groundwater overdraft. In contrast, some projects developed in California are controversial for their environmental implications, such as the State Water Project (SWP) and the Central Valley Project (CVP). The SWP moves water from water-abundant regions (the Feather River of the Sierra Nevada) to water-scarce regions (the west side of the San Joaquin Valley, the central coast and southern of California). This project is controversial because it has affected the aquatic ecosystem of the Sacramento-San Joaquin Delta by reversing the flows when water is moved from the Sacramento to Southern California through pumps and aqueducts. In addition, it made Southern California dependent of water from the north and more vulnerable to droughts happening in the north part of California. The SWP has also created a sentiment of resentment between people of the north who sees "their" water moved to the south and people from Southern California defending and securing "their" water resources in the north. The SWP and the CVP are similar to the construction of the Cutzamala system (Bunge et al. 2012), which generated common social and environmental problems of megaprojects (Caire Martínez 2005).

At a national level, important laws have been developed to protect and restore water resources in Mexico. Such is the case of the General Law of Ecological Balance and Environmental Protection (LGEEPA 2012) and the National Water Law (LAN 2016) stablished in 1988 and 1992, respectively. For climate change, the General Law for Climate Change (LGCC 2015) stablished in 2012 provides the framework for policies related to this issue. Other programs are developed at the beginning of each government administration, the National Water Program, the National Development Plan, and the National Infrastructure Program (CONAGUA 2014; PND 2013; PNI 2014).

However, there are challenges associated with the legal and institutional framework described above. First, there is a lack of execution and enforcement of the regulations mentioned above: it exists, but there is almost no enforcement in its application. Second, there is a lack of continuity in such policies; at the beginning of each presidential term, a series of plans are developed and then dismantled or redesigned all over again in the next presidential term. Thus, there is a lack of long-term planning as each administration lasts 6 years.

In terms of research, there are highly qualified scientists doing research in climate change and water resources; however, there is still a need for more applied research that can solve on-the-ground problems. In addition, there is no bridge between science (scientist) and policy design (decision-makers). Scientific projects may be funded but developed in vacuum without decision-maker's feedback and vice versa. Decisions are not made based on scientific results frequently funded and encouraged by policy-makers themselves. This is a chronic and systematic problem that has delayed or prevented the selection and implementation of scientific-supported solutions. Benefits from this system's integration and the understanding of human and hydrologic systems as a whole are essential to avoid fragmented science, management, and policy.

Some of the challenges of the suggested bottom-up approach is the selection of a diverse group of stakeholders that represent the different interests in the basin, which ultimately requires transparency in the selection process and a selection system that is based in the merits of each individual. The proposed system does not work in political environments that are biased by individual or institutional interest.

6.4 Conclusions

Extensive information by HAR is accumulated every year by CONAGUA. The information includes water use by sector, water quality on main rivers and water bodies, infrastructure, storage capacity, water stress, population with access to potable water and sewages, and groundwater extraction and aquifer conditions. However, this information is not integrated into a comprehensive analysis to address specific problems and propose a set of solutions for each of the HAR. Thus, there is a need for an IWRM modeling framework that integrates all the individual pieces into a system's dynamic model, which may include hydrologic, water allocation and system's operation, and environmental and social model components for every HAR.

Climate change information is accessible for every HAR; however, this information has not been translated into impacts on the ground in terms of (a) increased severity and frequency of droughts and change of water availability, (b) shifts in start and ending of rain seasons, (c) modification in agricultural growing seasons and increases/reductions in crop water needs, (d) increased magnitude and frequency of large rainfall events and related floods, (e) diminishing water quality due to pattern water cycle alteration, and (f) alteration/modification of habitat for ecosystems. Thus, it is necessary to evaluate the impact of climate change through water resources modeling and monitoring, as well as designing adaptive strategies to cope with climate change impacts.

In addition to carrying out new research, it is important to design institutional structures than can cope effectively with climate change impacts. Such institutions must be coupled with economic strategies and incentives to mitigate and adapt for changing climate conditions. Lastly, it is crucial to develop educational programs and materials that communicate the basics of climate change, current impacts on water resources, and actions to mitigate negative effects at the local, regional, and national levels.

Adaptive management strategies, review and formulation of new policies and regulations, and educational programs and incentives are needed at different institutional levels to successfully develop an IWRM framework (Hanak and Lund 2012). However, adaptation policies need to give particular attention to vulnerable populations, for which adaption may come at expense of other aspects of human welfare (Eakin et al. 2016).

Acknowledgments This material is based upon work supported by the UC-Mexico Initiative. The views and conclusions contained in this document are those of the authors and should not be interpreted as representing the opinions or policies of the involved institutions.

References

Bunge V, Martínez J, Ruiz-Bedolla K (2012) Caracterización y escenarios de dinámica hídrica de la región de aporte del sistema Cutzamala INECC México

Caire Martínez G (2005) Conflictos por el agua en la Cuenca Lerma-Chapala, 1996–2002. Región y sociedad 17:73–125

CCA (2016) Situación y contexto de la problemática del agua en México. Consejo Consultivo del Agua. http://www.aguas.org.mx/sitio/index.php/panorama-del-agua/diagnosticos-del-agua. Accessed Sept 2016

CONABIO (2016) Mexican biodiversity. National Commission for the Knowledge and Use of Biodiversity. http://www.biodiversidad.gob.mx/v_ingles/. Accessed September 2016

CONAGUA (2014) Programa Nacional Hídrico 2014–2018. Mexico D.F.

CONAGUA (2015) Statistics on water in Mexico 2015. Ministry of the Environment and Natural Resources. Mexico D.F.

CONAGUA (2016) Ley de aguas nacionales y su reglamento. National Water Commission (Mexico), Mexico City

De la Peña ME, Ducci J, Zamora V (2013) Tratamiento de Aguas Residuales en Mexico vol IDB-TN-521

DOF (2006) Acuerdo por el que se da a conocer el estudio técnico de los recursos hídricos del área geográfica Lerma-Chapala. Diario Oficial de la Federación,

Eakin H et al (2016) Adapting to risk and perpetuating poverty: Household's strategies for managing flood risk and water scarcity in Mexico City. Environ Sci Pol. https://doi.org/10.1016/j.envsci.2016.06.006

FAO-SAGARPA (2014) México: el sector agropecuario ante el desafío del cambio climático. Food and Agriculture Organization of the United Nations & Secretariat of Agriculture, Livestock, Rural Development, Fisheries and Food of Mexico

Feng S, Krueger AB, Oppenheimer M (2010) Linkages among climate change, crop yields and Mexico–US cross-border migration. Proc Natl Acad Sci 107:14257–14262. https://doi.org/10.1073/pnas.1002632107

Galindo LM, Reyes O, Alatorre JE (2015) Climate change, irrigation and agricultural activities in Mexico: a Ricardian analysis with panel data. J Dev Agric Econ 7:262–273. https://doi.org/10.5897/JDAE2015.0650

GWP (2000) Towards water security: a framework for action. Global Water Partnership, Stockholm

Hanak E, Lund JR (2012) Adapting California's water management to climate change. Clim Chang 111:17–44. https://doi.org/10.1007/s10584-011-0241-3

Herrera-Pantoja M, Hiscock KM (2015) Projected impacts of climate change on water availability indicators in a semi-arid region of central Mexico. Environ Sci Pol 54:81–89. https://doi.org/10.1016/j.envsci.2015.06.020
Hunter LM, Murray S, Riosmena F (2013) Rainfall patterns and U.S migration from rural Mexico. Int Migr Rev 47:874–909. https://doi.org/10.1111/imre.12051
IBWC (2012) Fact sheet: minute 319 environmental flow. United States and Mexico
INEGI (2016a) Ocupación y Empleo: Indicadores de ocupación y empleo al segundo trimestre de 2016. Instituto Nacional de Estadística Geografía e Informática. http://www3.inegi.org.mx/sistemas/temas/default.aspx?s=est&c=25433&t=1. Accessed Sept 2016
INEGI (2016b) PIB y Cuentas Nacionales: Producto Interno Bruto (PIB) – Trimestral. Instituto Nacional de Estadística, Geografía e Informatica. http://www.inegi.org.mx/est/contenidos/proyectos/cn/pibt/. Accessed Sept 2016
IPCC (2014) Climate Change 2014: synthesis report. Contribution of Working Groups I, II and III to the Fifth Assessment Report of the Intergovernmental Panel on Climate Change [Core Writing Team, R.K. Pachauri and L.A. Meyer (eds.)]. Intergovernmental Panel on Climate Change, Geneva
Jessoe K, Manning DT, Taylor JE (2018) Climate change and labour allocation in rural Mexico: evidence from annual fluctuations in weather. Econ J 128:230–261. https://doi.org/10.1111/ecoj.12448
LAN (2016) Ley General de Aguas Nacionales. Camara de Diputados del H. Congreso de la Unión
LGCC (2015) Ley General de Cambio Climático. Camara de Diputados del H. Congreso de la Unión
LGEEPA (2012) Ley General del Equilibrio Ecológico y la Protección al Ambiente. Camara de Diputados del H. Congreso de la Unión
Lopez-Feldman A (2013) Climate change, agriculture and poverty: a household level analysis for rural Mexico. Econ Bull 81:187–203
Loucks DP, Van Beek E, Stedinger JR, Dijkman JPM, Villars MT (2005) Water resources systems planning and management; an introduction to methods, models and applications. United Nations Educational, Scientific and Cultural Organization, Delf
Mendelsohn R, Arellano-Gonzalez J, Christensen P (2010) A Ricardian analysis of Mexican farms. Environ Dev Econ 15:153–171. https://doi.org/10.1017/S1355770x09990143
Monterroso Rivas AI, Romero JD, Pérez Espejo R, Martínez Córdova C, Márquez Rosano C, del Moral JB (2015) Sistemas de Producción de Alimentos y Seguridad Alimentaria. In: Gay C, García C, Rueda Abad JC (eds) Reporte Mexicano de Cambio Climático Grupo II: Impactos, vulnerabilidad y adaptación, pp 97–121
NMX-AA-159-SCFI-2012 (2012) Norm that establishes the procedure for environmental flow determination in hydrological basins. Diario Oficial de la Federación.
Ortiz-Partida JP, Lane BA, Sandoval-Solis S (2016) Economic effects of a reservoir re-operation policy in the Rio Grande/Bravo for integrated human and environmental water management. Journal of Hydrology: Regional Studies 1(8):130–144. https://doi.org/10.1016/j.ejrh.2016.08.004
PND (2013) Plan Nacional de Desarollo 2013–2018 (National Development Plan on Mexico). Mexico City
PNI (2014) Programa Nacional de Infraestructura (National Program of infrastructure of Mexico). Mexico City
PVWMA (2013) 2012 Basin management plan update. Pajaro Valley Water Management Agency, United States, CA, Watsonville.
Robles E, Flores L, Martìnez B, de Guadalupe Sáinz M, Durán Á, Ramírez E, González ME (2017) Water quality of houses in 13 municipalities in the state of Mexico. Science 2:350–353
Yunez-Naude A, Rojas-Castro LG (2008) Perspectivas de la agricultura ante reducciones en la disponibilidad de agua para riego: Un enfoque equilibrio general. In: Guerrero García Rojas HR, Yunez-Naude A, Medellín-Azuara J (eds) El agua en México: Consecuencias de las políticas de intervención en el sector. Fondo de Cultura Económica, México, D.F.

Chapter 7
The Transboundary Paso del Norte Region

Stakeholders' Preferences Allowing Water Resources Adaptation

Luzma Fabiola Nava

Abstract This chapter illustrates the potential to advance transboundary water resources management in a more comprehensive approach. The focus is given to the transboundary Paso del Norte (PdN) region which is considered as the most environmentally damaged, hydrologically developed, and prolific irrigation area in the Rio Grande/Bravo Basin. Stakeholders from the US-Mexico PdN region provide insights into what needs to be done to foster sustainable adaptation of water allocation and management. A preliminary set of policy recommendations aims to highlight stakeholders' preferences and interests and their integration into regional water resources management.

Keywords Paso del Norte region · Rio Grande/Bravo Basin · Transboundary region · Water allocation and management · Stakeholders · Adaptation

7.1 Introduction

The USA and Mexico share a nearly 3200-km-long border that crosses three river basins: the Colorado River (CR), the Tijuana River (TR), and the Rio Grande/Bravo (RGB). This paper focuses on a significant RGB transboundary region: the surface water resources of the Paso del Norte (PdN) region (hereinafter the term water refers to surface water resources unless otherwise specified). The PdN region is located right at the midpoint of the US-Mexico border. The region extends approximately 547 km along the RGB from Elephant Butte Reservoir in southern New Mexico (NM) to the confluence of the Rio Conchos in Presidio County, Texas (Tx). The water of the PdN is allocated within a binational legal framework that allocates water resources, sets extractions and diversions, and sustains political boundaries

L. F. Nava (✉)
Center for Global Change and Sustainability, C.A. (CCGS), Villahermosa, Tabasco, Mexico

International Institute for Applied Systems Analysis (IIASA), Laxenburg, Austria
e-mail: luzma.nava@ccgs.mx

E. O. Vieira et al. (eds.), *Integrated Water Resource Management*,
https://doi.org/10.1007/978-3-030-16565-9_7

between two federal countries, three federated states, and three metropolitan areas. Allocations to each side of the border are based on an estimate of the RGB Basin's hydrology at the time decisions were made (1900–1944). These have remained unchanged despite having on average 17% less water than expected after the treaty was signed (1950–2004) (CONAGUA 2008; R. J. Brandes Company 2004) and despite a substantial increase in municipal, industrial, and agricultural water demands, multiple coexisting water users, and unmet environmental and sustainability needs. Limited water resources in the context of an arid climate and related ecological vulnerabilities invite competition among water users and water stakeholders across and along the border, taxing the quality of the river environment and human capacity to solve common problems.

As stated in Nava et al. (2016), US-Mexico shared water challenges have been successfully addressed on an ad hoc basis in two of the transboundary river basins. Binational agreements on the CR and the TR have fostered water management adaptation through ecological restoration, qualitative riparian improvements, and stakeholder involvement. The creation of binational groups of competing stakeholders is one of the key demarches in implementing new legal instruments to solve common river basin issues. But these developments have been slow to emerge in the RGB. The RGB has been managed largely to meet competing binational demands driven by regional economics, with little regard of the river's instream flow (Nava et al. 2016). Within the RGB, the PdN region is the most obvious example of (1) the stressed water allocation system, (2) the difficulty of reconciling water uses for coupled economic and environmental purposes, and (3) the binational challenges imposed by securing water in the border and accommodating stakeholders' preferences and interests.

Based on stakeholders from the PdN region, this chapter answers the following questions: (1) How can the US-Mexico water allocation framework be adapted to support and preserve the RGB Basin ecosystem? (2) What are the main opportunities to be explored for fostering sustainable adaptation of water allocation and water management in the PdN? (3) How stakeholders' preferences and interests can translate into policy recommendations? Posing these questions directs attention to the existing opportunities to adapt the existing PdN regional water resources management through stakeholders' participation and reinforce the cooperation over transboundary water resources.

7.2 Background Information

The RGB is the longest of the three shared rivers between the USA and Mexico and forms their international boundary for 2034 km (Fig. 7.1).

The Rio Grande, as it is known in the USA, or the Rio Bravo, as it is called in Mexico, is the 20th largest river in the world, with a total length of 3059 km. Its headwaters are located in the state of Colorado in the USA. The river flows south through New Mexico (NM) arriving well south of Albuquerque at two major

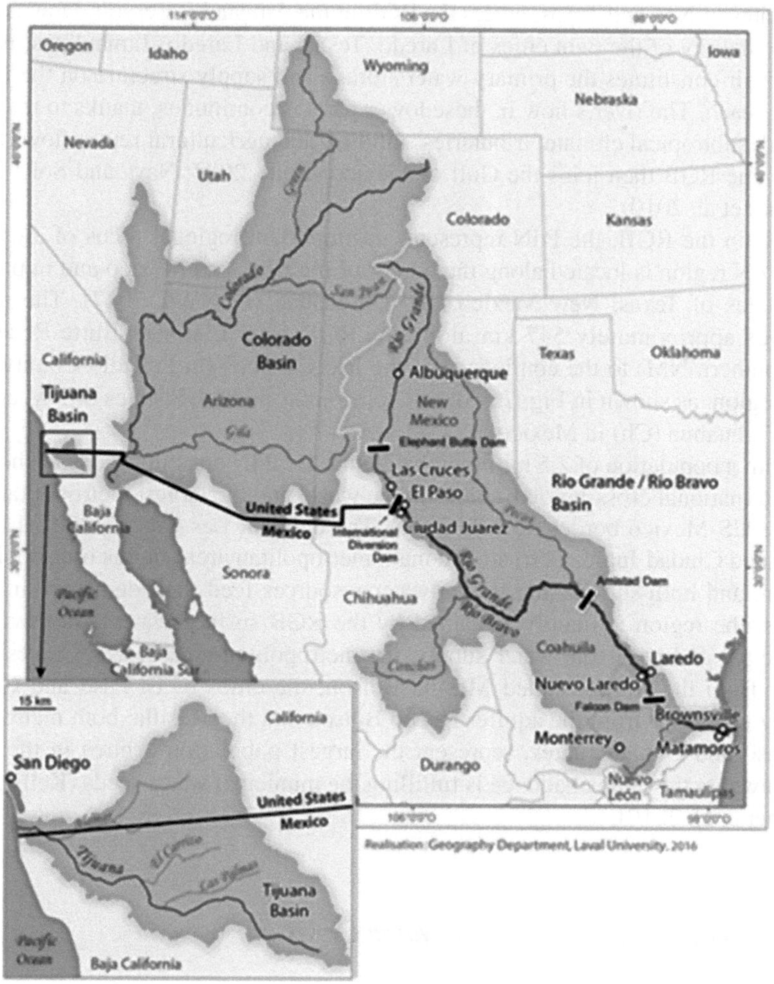

Fig. 7.1 The US and Mexico transboundary rivers (Nava et al. 2016)

reservoirs on its course, Elephant Butte (EB) and Caballo dams. Downstream of these, the flow almost entirely depends on releases from the reservoirs. Starting from the cities of Ciudad Juarez (Chihuahua, Mexico) and El Paso (Texas, USA) until its mouth in the Gulf of Mexico, the RGB forms the international border between Texas and the Mexican states of Chihuahua, Coahuila, Nuevo León, and Tamaulipas. Due to the semiarid climate and the extensive withdrawal of water for irrigation, the flow of the river in the section downstream of Ciudad Juarez/El Paso is severely depleted and is aptly called the Forgotten Reach. The Rio Conchos, in the Mexican state of Chihuahua, replenishes the RGB which then flows through the ecologically unique, mountainous area of the Big Bend Reach, protected in the form of national parks and natural protected areas. The Pecos River, flowing through the

US states of NM and Texas, enters the RGB at the Amistad Reservoir. Downstream in the vicinity of the twin cities of Laredo, Texas, and Laredo, Tamaulipas, Falcon Reservoir constitutes the primary water storage and supply structure on the river's lower reach. The river's flow in these lower parts is continuous, thanks to the more humid, subtropical climate, tributaries' inflows, and agricultural return flows to the river. The RGB then joins the Gulf of Mexico (Kelly 2002; Nava and Solis 2014; Parcher et al. 2010).

Within the RGB, the PdN represents an important regional focus of attention. The PdN region is located along the border of the USA and Mexico and in parts of the states of Texas, New Mexico, and Chihuahua (PDNWC 2017). The region extends approximately 547 km along the RGB from Elephant Butte Reservoir (in southern NM) to the confluence of the Rio Conchos (in Presidio County, Tx). The region, as shown in Fig. 7.2, drains some areas in the US states of NM and Tx and Chihuahua (Ch) in Mexico.

With a population of 2.5 million inhabitants, the PdN constitutes one of the largest international cross-border regions in the world and the largest metropolitan area on the US-Mexico border (OECD 2010). The cities of Las Cruces (NM), El Paso (Tx), and Ciudad Juarez (Ch) are the main metropolitan areas in this transboundary region, and both surface and groundwater resources feed their demands. In other words, the region is mainly sustained by the RGB surface waters. However, for municipal and industrial water supply, the metropolitan area of Las Cruces takes water from the aquifer called Mesilla Bolson; the cities of El Paso and Ciudad Juarez use water from the aquifer Hueco Bolson and the Mesilla; both metropolis, El Paso and Ciudad Juarez, represent the largest population centres in the RGB Basin where the main challenge is fulfilling the municipal water needs (Kelly 2002; Parcher et al. 2010).

7.2.1 Quantity and Quality Water Crisis

The PdN region suffers from being a metropolitan transboundary region. In the US-Mexico RGB Basin, the PdN is the most irrigation-intensive and environmentally damaged area (Riley 1995). With an average annual rainfall of about 200 mm of rain and increasing evaporation rates, crops such as maize, alfalfa, pecans, chilli, and cotton are growing in this semiarid region. Climate change is expected to have consequences in the water cycle by reducing surface flows and aquifer recharge rates. It is also expected to affect the quality of aquatic environments and accentuate sustained drought and water scarcity. This regional environmental damage translates into water quantity and water quality issues. The RGB Basin personality translates into a pauperized river system where water is scarce, and low precipitation levels lead to a critical limiting factor for development. In the RGB, the limited availability of water restricts environmental management capacity. However, while the headwaters are fed by snowmelt from the San Juan Mountains in Colorado, most

Fig. 7.2 The Paso del Norte region (available online at http://www.pdnwc.org/)

of the basin is located in the Chihuahuan Desert, with an average precipitation of less than 500 mm (PRISM Climate Group 2016). From this limited supply, significant withdrawals for human and economic utilizations are made basin wide despite the fact that water quantity problems may be exacerbated by climate change. As it is well known, water quantity problems amplify the existing issues of water quality and pollution. Water quality degradation is imperilling and stressing the delicate ecosystems supported by the river. Issues such as salinity, bacteria, ammonia, chlorophyll, and nitrate, residual pharmaceuticals in wastewater, and lack of green corridors and loss of biodiversity are accentuating impoverishment of natural ecosystems (Nava et al. 2016).

This context is increasing the likelihood of a water crisis in the PdN region. The PdN region continues to experience rapid population growth despite its diminishing water resources. Institutional fragmentation and hydrological organization in this – binational and tri-state and metropolitan – region prevents water managers' preparedness for upcoming climate challenges as well as the development of adaptive management strategies (Nava and Solis 2014). Nevertheless, stakeholders from the transboundary Paso del Norte may have the potential to foster sustainable adaptation of water allocation and water management and therefore to update and provide feedback on the design of binational water transboundary policies.

7.2.2 *The Binational Instruments Shaping the Transboundary Water Allocation Framework*

The RGB water allocation binational framework (Fig. 7.3) is based on international agreements between the USA and Mexico and compacts at the national level among the US riparian states (Nava et al. 2016; Nava 2017; Nava and Solis 2014).

For the purpose of this chapter, the focus is given to the treaty between the USA and the United Mexican States relating to the utilization of the waters of the Colorado and Tijuana rivers and of the Rio Grande (Rio Bravo) known widely as the 1944 Water Treaty (IBWC 1944). The 1944 Treaty aims to obtain the most complete and satisfactory utilization of shared waters based on the equitable distribution between the two countries of the waters of shared river systems. Under the 1944 Treaty, the International Boundary Commission (IBC), established in 1889 to

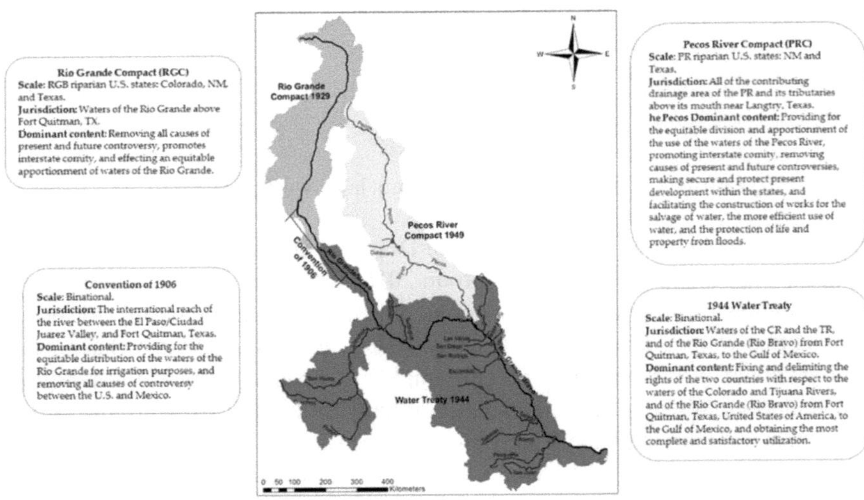

Fig. 7.3 The RGB Basin binational instruments for water allocation (Nava et al. 2016)

rectify and maintain the riparian boundary of the Rio Grande and Colorado rivers, became the International Boundary and Water Commission (IBWC, hereinafter called the Commission), with specific mandates on shared water resources. The 1944 Water Treaty establishes the Commission's authority as an international body and endows it with considerable flexibility in addressing existing and emerging issues through the *minute process* (Nava 2017).

A *minute* technically represents "the written record of meetings, particularly of [...] Stakeholders [...]" (West's Encyclopedia of American Law 2015). As such, *minutes* are extensions and applications of the treaties and offer the potential for adapting the US-Mexico water allocation framework to changing circumstances (Murcia 2013). Under common treaty law, an amendment is a new treaty and requires ratification (McCaffrey 2006). However, Article 25 of the 1944 Treaty provides for this flexible procedure that allows the two countries to adapt their boundary and water treaties to new circumstances (IBWC 1944). In all, 323 minutes have been signed since the 1944 Water Treaty entered into force in November 1945. Of these 140, a total of 71 concern the RGB Basin (Nava et al. 2016). From which, the most relevant to the issue that I am dealing with here concerns the Minute 234, 1969; Minute 293, 1995; Minute 307, 2001; Minute 308, 2002; and Minute 309, 2003 (Mumme 2010). These minutes are mainly focused on the environmental governance in the RGB and the mechanisms dealing with transboundary water issues. The binational *minute process* translates into a flexible and adaptable 1944 Water Treaty ensuring the legal capacity for addressing changing circumstances and supporting continuing and emerging issues not explicitly included in the Treaty.

7.2.3 *Qualitative Methods*

This research employs an interdisciplinary and qualitative approach to problem identification and analysis. Our qualitative approach is composed of various methodological tools, including case study analysis, collection of documents, and field work and semi-structured interviews. The PdN round of field work consists on the application of a questionnaire (40 multiple-choice questions and 10 open-ended questions) to gain information on water resources management and sustainable practices. A total of 23 interviews were conducted between October and November 2015. Taking into account the location of the respondents, the questionnaire was applied on the basis of their availability, in places that respondents preferred, either their workplace or a public place. In situations where travel was difficult, the questionnaire was sent by the Internet. In some cases, answers to open-ended questions were recorded with a digital recorder and transcribed with NVivo 9 and NVivo 10 software. We used NVivo 10 to do qualitative content analysis (Nava 2017; Nava et al. 2016). For ethical reasons, we ensure the confidentiality of the participants. However, more specific details on the stakeholders' profile and questionnaire process can be found in Nava et al. (2016).

7.3 A Look Through the PdN Region Stakeholders' Insights to Allow Water Resources Adaptation

All those questioned in this assessment agreed on the importance of enhanced water resources management to deal with vulnerability in an arid context. Stakeholders' insights have been systematized in three solutions-options reflecting their main visions to solve common problems and foster sustainable water resources adaptation in the PdN region.

7.3.1 How Can the US-Mexico Water Allocation Framework Be Adapted to Support and Preserve the RGB Basin Ecosystem?

Stakeholders from the PdN region place great emphasis on the need to (a) strengthen communication and articulation among all of stakeholders in the RGB Basin and related water agencies, (b) provide environmental education, (c) manage surface water and groundwater jointly, and (d) renegotiate all river basin water agreements due to imbalances between water availability, supply, and demand. In order to achieve these goals, stakeholders from the PdN region highlight the challenges and the opportunities related to living in an arid area and desert landscape. The challenges are related to the given environmental conditions. Water in a desert environment is not an abundant available resource. What is abundant is the determination and the capacities that plants, animals, and people living in the area have deployed to adapt and survive. The opportunities in the PdN region are related to the development of unique customs, procedures, and technologies to address difficulties in this region.

7.3.2 What Are the Main Opportunities to Be Explored for Fostering Sustainable Adaptation of Water Allocation and Management in the PdN?

Stakeholders from the PdN region recognize the *minute process* as the main platform – and maybe the unique binational instrument – for framing stakeholders' concerns and addressing water quantity and quality issues. A member of the US Section of the Commission explains the procedure: a *minute* is the result of top-down and bottom-up decisions approved by the two governments through the Commission. This respondent also highlights that the *minute process* is a challenging dynamic process that could last many months or years. However, each *minute* is a unique agreement based on the identification of a problem or a need along the US-Mexico border that is within the authorities of the IBWC. To create a *minute*, the Commission has to engage in a binational dialogue in order to collaborate in technical or engineering studies or reviews about the problem or the need. Based on those technical studies or reviews, the Commission elaborates a series of recommendations. The Commission then drafts the recommendation in a *minute* format. Engineering and diplomatic staffs of the Commission play an important role in the conception of the *minute*, as they initiate the consulting process with key stakeholders involved in the recommendations that will be proposed. Finally, *minute* drafts are sent back and forth in English and Spanish to the US and Mexican sections of the Commission after being informed by key stakeholders' comments and agreed by the parties. The *minute* is then submitted for approval to the US Department of State and the Mexican Ministry of Foreign Relations. Once the *minute* is approved, it becomes a binding agreement in the two countries (Nava et al. 2016). In the same order of ideas, some other PdN stakeholders recognize that, due to potential risks of conflicts over water supplies and access to water, the *minute process* is the most reliable political instrument to frame regional and binational concerns but more specially to consider their concerns and interests regarding the use, allocation, and preservation of water in the region.

7.3.3 How Stakeholders' Preferences and Interests Can Be Translated into Policy Recommendations?

At the moment, the only binational mechanism available to solve a shared water-related problem and allow water resources adaptation in the PdN region is the *minute process*. However, it may be one additional means to make sure voices of stakeholders are heard.

At the RGB Basin scale, stakeholders highlight the importance to create an exploratory RGB Basin Task Force to obtain recommendations from river basin stakeholders, such as in the Tijuana River Basin, the Lower Colorado River, and the Colorado River Delta (Minutes 317–320). At the regional level, stakeholders from the PdN recognize the value to relive what once was the *Paso del Norte Water Task Force*.

The *Paso del Norte Water Task Force (PdNWTF)*, thought in 1997, aimed to reconcile the region's agricultural and municipal interests (Hamlyn 2001). It also aimed to assemble a broad-based group of stakeholders working cooperatively to promote a binational and tri-state perspective on water issues impacting the future prosperity and long-term sustainability of the PdN region (PNWTF 2002). But sadly, lack of funds has prevented the PdN group of stakeholders from ensuring a solid and lasting regional and transboundary cooperation. However, some members of this group have recently taken the initiative to rescue and resuscitate the PdNWTF efforts.

7.4 A Preliminary Set of Binational and Regional Policy Recommendations

The concluding section illustrates a preliminary set of policy recommendations highlighting stakeholders' preferences and interests and their integration into regional water resources management. One of the most recent activities from the reignited PdNWTF is to conduct a survey on water scarcity in the region. The results from the Survey of the Paso del Norte Water Stakeholders (Kibaroğlu and Schmandt 2016) show important insights into what stakeholders consider a means to foster sustainable adaptation of water allocation and water management in the region. As it has been mentioned, drought in the PdN region has an important impact on water plans. Drought affects water availability for all water users, restraints economic and social benefits, and makes the future of the region less predictable. As an immediate solution to regional drought impacts, participants on this survey put greater emphasis on water conservation, increased groundwater pumping, and efficient irrigation technologies. In this regard, the results of this survey could be summarized and taken as the starting point for the elaboration of binational and regional policy recommendations. Figure 7.4 shows some of the most important insights from stakeholders to enable the sustainable adaptation of water allocation and water management needs to be done.

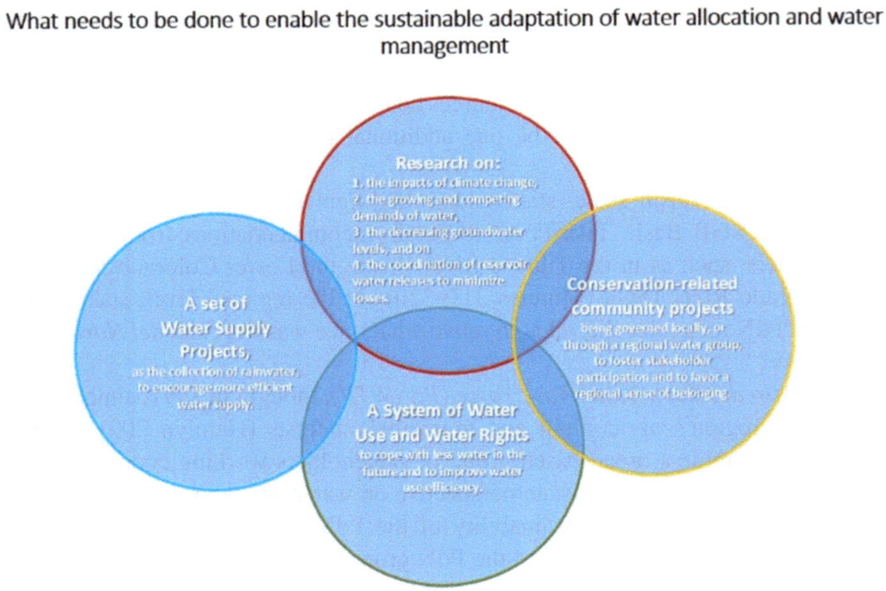

Fig. 7.4 Insights from stakeholders. Prepared by the author based on the results of the PdN survey from (Kibaroğlu and Schmandt 2016)

The success of sustainable adaptation of water allocation and water management is directly linked to the effectiveness of stakeholders' collective action taken at the regional level. Having said this, for the PdNWTF to succeed, it is important to achieve acceptance and recognition by both already established water authorities, water agencies, and existing water users. To advance transboundary water resources management in a more comprehensive approach, all stakeholders from the PdN are encouraged to continue sharing their ideas and implementation experiences for others to learn from. And in this context, the PdNWTF would represent the entity enabled to capture the dynamic and the evolving dimensions of water-related issues and to offer a holistic approach for addressing water challenges through the umbrella of sustainable adaptation of water allocation and water management (UN-Water 2013).

To conclude, the set of stakeholders' insights represents an ensemble of regional recommendations, which, if framed into the *minute process*, could translate into an extension of the 1944 Water Treaty, an extension that would potentialize the sustainable adaptation of water allocation and water management in the Paso del Norte region. Hopefully, the PdNWTF would be able to muster all its forces in order to commence meaningful deliberations on the substantive regional water issues, so as to demonstrate its continuing potential as an important, indeed vital, binational group of competing stakeholders for the discussion of the regional sustainable water management adaptation. The PdNWTF could become a regional magnet and institution for water cooperation with the aim to promote the implementation of the regional programmes based on stakeholders' insights.

Acknowledgements The author would like to thank all anonymous stakeholders and experts for their willingness to participate in the interviews by providing meaningful insights.

References

CONAGUA (2008) Acuerdo por el que se da a conocer el resultado de los estudios de disponibilidad media anual de las aguas superficiales de la cuenca del Rio Bravo. Diario Oficial de la Federación, México City

Hamlyn E (2001) Paso del Norte Sustainable Water Use Strategy. In D.E. Stott, R.H. Mohtar and G.C. Steinhardt (eds). Sustaining the Global Farm. Selected papers from the 10th International Soil Conservation Organization Meeting, May 24-29, 1999, at Purdue University and the USDA-ARS National Soil Erosion Research Laboratory, pages 090–094

IBWC (1944) Treaty for the utilization of waters of the Colorado and Tijuana Rivers and of the Rio Grande. International Boundary and Water Commission, Washington, DC

Kelly ME (2002) Water management in the Binational Texas/Mexico Rio Grande/Rio Bravo Basin. Yale For Environ Stud Bull 115:134

Kibaroğlu A and Schmandt J (2016) Sustainability of Engineered Rivers in Arid Lands: Euphrates-Tigris and Rio Grande/Bravo, The LBJ School of Public Affairs, The University of Texas at Austin. ISBN: 978-0-89940-818-7 https://repositories.lib.utexas.edu/handle/2152/39395 Accessed 1 July 2017

McCaffrey SC (2006) Understanding International Law; Elsevier: LexisNexis.

Mumme S (2010) Environmental governance in the Rio Grande watershed: binational institutions and the transboundary water crisis. J Transbound Water Resour 1:43–68

Milanés-Murcia M (2013) A New International Legal and Institutional Framework to Manage Fossil Aquifers and Groundwater in Conjunctive Use with Surface Water along the U.S.-Mexico Border: A Water Banking Perspective; University of the Pacific, McGeorge School of Law Dissertations: Sacramento, CA, USA. https://scholarlycommons.pacific.edu/mcgeorge-dissertations/1 Accessed 1 July 2017

Nava LF (2017) Water security through institutional resilience in the transboundary Paso del Norte region. In: World environmental and water resources congress 2017, pp 464–475

Nava LF, Sandoval-Solis S (2014) Multi-tiered governance of the Rio Grande/Bravo Basin. Int J Water Gov 2:85–106

Nava LF, Brown C, Demeter K, Lasserre F, Milanés-Murcia M, Mumme S, Sandoval-Solis S (2016) Existing opportunities to adapt the Rio Grande/Bravo Basin Water Resources Allocation Framework. Water 8:291

OECD (2010) Higher education in regional development: Paso del Norte region, Mexico and the United States. Organisation for Economic Cooperation and Development. https://www.oecd.org/mexico/45820961.pdf. Accessed 25 July 2016

Parcher JW, WooDWarD DG, Durall R (2010) A descriptive overview of the Rio Grande--Rio Bravo watershed. J Transbound Water Resour 1:159–178

PDNWC (2017) Our watershed. Paso del Norte Watershed Council. http://smiley.nmsu.edu/pdnwc/watershed.html. Accessed 21 Nov 2017

PNWTF (2002) Mission. Paso del Norte Water Task Force. http://www.meadowscenter.txstate.edu/rg/database_profile.php?iid=31. Accessed 21 Aug 2017

PRISM Climate Group (2016) Annual average precipitation of the contiguous United States. Northwest Alliance for Computational Science & Engineering. http://www.prism.oregonstate.edu. Accessed 16 May 2016

R. J. Brandes Company (2004) Final report – water availability modeling for the Rio Grande Basin. Texas Commission for Environmental Quality (TCEQ)

Riley CL (1995) Rio del Norte: people of the upper Rio Grande from earliest times to the Pueblo revolt. University of Utah Press, Salt Lake City

UN-Water (2013) Water security & the global water agenda UN Water analytical brief. UN University, Hamilton

West's Encyclopedia of American Law (2015) e.S.v. Minutes. http://legal-dictionary.thefreedictionary.com/Minutes. Accessed 19 Nov 2015

Chapter 8
Water Governance and Adaptation to Drought in Guanacaste, Costa Rica

Ricardo Morataya-Montenegro and Pável Bautista-Solís

Abstract In this chapter we review the key learnings and challenges for water management in a territory where water is severely affected by climatic variability: the Guanacaste province in Northwestern Costa Rica. In this territory the water governance system is contested by the interaction of biophysical, cultural, and political factors, creating conditions for the emergence of disputes and enhancing the environmental and economic externalities from economic activities, mainly agriculture and tourism. We review the main factors from these intertwined dynamics to provide key lessons and identify sensible gaps in knowledge that need to be addressed in the upcoming research and integrated water resource management efforts. Our work shows that climate variability is increasing water demand, calling for a contextualized policy for managing water in Guanacaste. Moreover, the centralized, vertical, and fragmented water governance system led by the Central Valley region is imposing challenges for building up an adaptive governance system aiming for resilience at a long temporal scale. Despite the latter, several community-led experiences facilitated by boundary organizations and local champions suggest that water in Guanacaste can be secured by establishing multi-sectoral platforms for water adaptive governance and increasing the decision-making based on technical and scientific information.

Keywords Drought · Water governance · Water conflicts · Costa Rica · Climatic variability

R. Morataya-Montenegro (✉)
Universidad Nacional, Heredia, Costa Rica
e-mail: morataya@una.ac.cr

P. Bautista-Solís
Mesoamerican Center of Sustainable Development of the Dry Tropics (CEMEDE),
Universidad Nacional, Heredia, Costa Rica
e-mail: pavel.bautista.solis@una.ac.cr

E. O. Vieira et al. (eds.), *Integrated Water Resource Management*,
https://doi.org/10.1007/978-3-030-16565-9_8

8.1 Starting Point: An Overview of Water Scarcity Issues in Guanacaste, Costa Rica

Costa Rica is a small country (Fig. 8.1) located in the Isthmus of Central America, between the Pacific Ocean and the Caribbean Sea, with a continental area of 51,100 km^2 and 589.683 km^2 of territorial sea. The country is located at a latitude between 8 and 12 degrees North of the equatorial line. Hence, its climate is humid tropical, with abundant rainfall on the Caribbean Coast and lowlands and quite the opposite in the North Pacific, where the province of Guanacaste is located (Costa Rica 2017).

The Guanacaste province shows a climatic variability featured along the whole Central American Dry Corridor. This is an irregular area facing seasonal aridity and recurrent droughts, which is located mainly in the Pacific watersheds from the Central American Isthmus (Hidalgo et al. 2019; Hidalgo et al. 2016). Here, a well-established seasonal aridity season contrasts with a generous rainy season. During the former, high temperatures (27–36 C° on average) and limited rain produce water scarcity from December to May, being the most critical months from March to April. The rainy season (annual mean of 2385 mm/year) is usually expected in late May and cease in November (Solano and Villalobos w.d.).

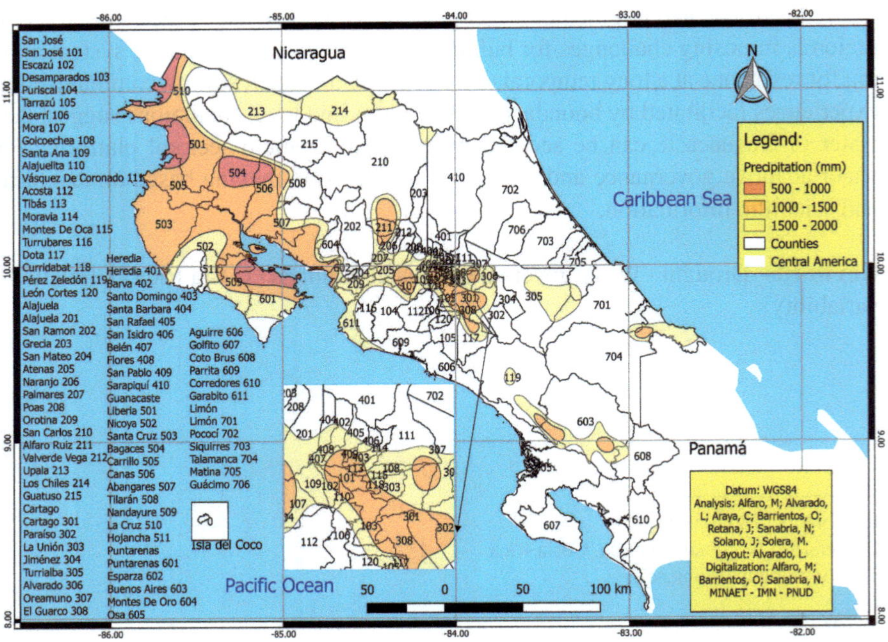

Fig. 8.1 Spatial location of the dry corridor of Costa Rica (Retana et al. 2012)

Several biophysical and anthropogenic factors are combined to aggravate water scarcity impacts in Guanacaste and the Central American Dry Corridor. Firstly, recurrent droughts are observed in an average of 10 years considering the dry corridor and 7 years in average in Guanacaste (Hidalgo et al. 2019; Retana et al. 2012). Secondly, many of the drought events are related to warm phases of ENSO, locally known as El Niño, although only eight of every ten El Niño events produce a drought in Costa Rica (Retana et al. 2012). Thirdly, other studies have showed a trend for an increased extension and severity of midsummer droughts, which are expressed as an interruption of the rainy season observed in mid-July (Hidalgo et al. 2013; Maurer et al. 2017). Fourthly, water demand has been increasing due to the demographic expansion (INEC 2012) and the growing demand of the coastal tourism industry and agricultural industry (i.e., sugarcane, melon, and watermelon). Fifthly, a weak water governance limits the coordinated implementation of an integrated management of water resources (Kuzdas and Wiek 2014; Kuzdas et al. 2015b). Finally, a culture of water abundance and a centralized strategic planning and decision-making always ignore all the latter aggravating conditions, hindering the implementation of customized actions for increasing water security in Guanacaste.

Local economic activities and domestic users struggle to obtain water at least during the six months of seasonal aridity. This is aggravated by a reactive strategic planning always focused on implementing solution alternatives after water security issues are already causing major impacts. For instance, there are several coastal aquifers with salt intrusion issues in the Santa Cruz County. This issue shall have been prevented at least by regulating water extraction to avoid the freshwater tipping point from such aquifers. Instead, extraction remained constant until water quality analysis did not conform the national standards, then the wells were closed, and new supply sources were searched. There are large water infrastructure projects such as the Program for Water Supply for the Middle Watershed from the Tempisque River and Coastal Communities (in Spanish, Proyecto de Abastecimiento de Agua para la Cuenca Media del Río Tempisque y Comunidades Costeras PAACUME), a large aqueduct taking water from the DRAT for expanding the irrigation district and transferring water to the coast of Santa Cruz, which have been discussed in the governmental portfolio for decades.

Finally, climate change is showing to have an important pressure on Guanacaste's water systems as it has been identified as a trend on increasing annual mean temperatures along the dry corridor. This is expected to reduce surface runoff in the region, increasing water demands for ecosystems and societal needs as evapotranspiration rates are increasing. Moreover, climate change is expected to increase the severity and frequency of extreme weather events. The Guanacaste province is impacted by droughts but also by floods, such impact may impose more challenges to water security, especially in the most rudimentary drinking water systems located in rural communities. However, more studies are needed to understand the implications of the latter in Guanacaste.

8.2 Observed Regional ENSO-Related Drought Impacts (2014–2016)

Before continuing, it is important to have a regional perspective about the impact of the ENSO-related drought in Central America. For this we present a summary of the latest event documented: the ENSO drought (2014–2016). According to the State of the Nation Program in 2016 (Programa Estado de la Nación 2010), during the rainy season of 2014 and 2015, the Pacific watersheds of Central America experienced a deficit of rainfall that caused moderate and severe drought conditions in seven countries, particularly in Central American Dry Corridor. This drought event caused negative impacts on ecosystems, agriculture, water resources, and food security. Precipitation records at several measuring stations reported the lowest cumulative rainfall minima of the last 40 years. According to the CRRH-SICA (2015), the 2014–2015 abnormally dry rainy season was a consequence of the effect produced by El Niño. Moreover, a cooling of the surface of the Atlantic Ocean, and especially the Caribbean Sea, generated a temperature difference in the seas on both sides of the Isthmus with a magnitude never observed before (CNE 2015). During the seasonal aridity from 2014 to 2015, there were locations in Central America where it did not rain for up to 42 consecutive days, and in some cases, the rainfall deficit amounted more than 65% of the historical average volume for the dry season.

The drought generated additional water rationing in some cities and rural municipalities; pumping costs were increased for the extraction of water from aquifers, or superficial sources were completely dried, as were used to supply small rural communities. In Guatemala, 16 of the 22 departments reported damages associated with the water deficit, as well as 10 of the 14 departments of El Salvador (CRRH-SICA 2015). According to the Obsan-R–SICA (2015), this phenomenon affected agriculture (mainly staple crops, i.e., maize and beans), especially subsistence activities in Guatemala, Honduras, El Salvador, and Nicaragua (Chen et al. 2018). This exacerbated nutritional and food security issues, especially for populations located in highly vulnerable areas such as many indigenous and rural communities. Although there was no food shortage at the regional level, the communities most affected by the drought were those where food and nutrition security are more fragile and where damage to basic grain crops was severe. According to the CRRH-SICA (2015), in some countries of the North of the region, whose inhabitants have a high dependence on maize and beans, the losses reached up to 80% and 60% of the harvests of those crops, respectively.

In Panama, the rice cultivation (especially rainfed rice) area of Chiriquí was the most affected. The coffee sector was also severely impacted, except for Nicaragua. According to Ramirez and Bonilla (2015), in Central America the drought caused damage in coffee plots of thousands of farmers; this was cumulative to the impact of the coffee leaf rust. The ENSO-related drought significantly reduced soil moisture, which hindered the absorption of nutrients. The total number of families affected in 2014 was at least 512,068. These families suffered losses in their means of production, their food sovereignty, and access to adequate water for their consumption. Regarding the livestock sector, the shortage of water for both animal and pasture had consequences on animal production and reproduction. In addition, the lack of rainfall increased pest incidence affecting both livestock and crops (Obsan-R–SICA 2015).

In the Central American Dry Corridor, conflicts between the population and the business sectors have arisen because of water scarcity. This is aggravated by the lack of water storage infrastructure. In the Panama Canal, the month of June 2014 was reported as the driest of the year; in the province of Guanacaste, Costa Rica, it experienced a long period of drought; and in Nicaragua, the dry season spread throughout the month of August 2014 (Obsan-R–SICA 2015).

Since April 2014, the national meteorological and hydrological services, articulated in the Climate Forum of Central America and coordinated by the Regional Committee on Hydraulic Resources (CRRH-SICA), issued alerts on the probability that in 2014, the rains in the dry corridor were below normal (Ramirez and Bonilla 2015). Updates about the persistence of the precipitation deficit were reported every three months during the years 2014 and 2015. This information prompted the governments of the Central American region to create committees and enact emergency decrees, among other actions to face the drought and mitigate their impact on the most vulnerable productive activities and population groups. Given its characteristics of slow progression and variable duration, drought requires medium- and long-term investments and approaches, as well as specific national policies and plans, as proposed by the UN, which has highlighted droughts as the most destructive extreme weather event affecting developing countries (UNCCD et al. 2013). In summary, the 2014–2016 ENSO-related drought is one of the most severe droughts ever observed, compared only with the drought events from 1982–1983 and from 1997–1998, impacting the regional food security of 3.5 million people (Sánchez-Murillo et al. 2017).

8.3 Costa Rica: Guanacaste and Its Environmental Dynamics

The environmental dynamics of Guanacaste are focused mainly on the presence or absence of forest cover (ecological indicator). It can be noted that historically it is one of the most important factors that have affected or benefited the provision of ecosystem services in this area. Land use in Guanacaste is the result of the economic policies of the Costa Rican State, where deforestation was promoted, arguing that farmers had to give value to their work plots by converting them into pastures for cattle or agricultural plots. Guanacaste was the agricultural frontier and was expected to supply food at lower prices for the rest of the country. Otherwise, farmers were not subject of credit and financial incentives to favor an international market that demanded beef in those years (approximately 40 years of the twentieth century). Later, the beef market in Central America fell, because Argentina and Brazil began to cover beef demand at a lower price but also because during the 1970s and 1980s, severe recurrent droughts impacted water availability (Caviedes 1997).

Fig. 8.2 Riparian forest cover from the Rio Lajas, Quiriman, Nicoya. (Guanacaste Picture: Ricardo Morataya-Montenegro)

The latter discouraged cattle raising in Costa Rica, farmers abandoned or sold their land and returned to their places of origin, which affected these families economically but at the same time favored the deforested land for cattle to recover in a succession process. -path to recuperate forests cover (Fig. 8.2).

Nowadays, the forest cover of Guanacaste is recuperated, especially in the Nicoya Peninsula. However, primary forest and their ecosystem services are lost as the remaining forest cover is comprised of silvicultural plantations, early stages of natural succession, and secondary forests, which emerged after the collapse of the price of beef and abandonment of land occurred. The recuperation of forest cover has been related to the emergence of another main economic activity for Guanacaste: tourism (Morataya-Montenegro 2011). This is because Costa Rica is a referent in such industry, receiving millions of tourists per year interested in enjoying the wilderness provided by conservation policies for favoring the environment (25% was reserved under a category of protection).

8.4 Water Governance in Guanacaste and Costa Rica

The growing awareness at the regional level on issues such as the unsustainable use of water, its pollution, its monopolization by important economic sectors, and the lack of efficiency and cover from the Latin American states to administrate this vital resource demonstrates the relevance of governance for an integrated water resources management (IWRM). In this work we embrace the theoretical advances on water governance generated by Claudia Pahl-Wostl (2015). This scholar reviewed three different approaches while framing governance: (i) *politics*, or the leverage for creating norms and legislation for the multiple and confronted interests of water stakeholders; (ii) *polity*, encompassing institutions and their rules to moderate the behavior of water stakeholders, i.e., suppliers, distributers, users, and disposers; and (iii) *policies*, as the different instruments for water regulation, including those considered nonformal, formal, and market-based. Considering the later water governance is defined as "… the social function that regulates development, and management of water resources and provisions of water services at different levels of society and guiding the resource towards a desirable state and away from an undesirable state" (Pahl-Wostl 2015).

In a small country such as Costa Rica, water and institutional normatives are similar across scales. However, there are some differences regarding their influence on water management and their power on policy-making across scales (Fig. 8.3). Moreover, local or regional nonformal and formal agreements are observed in Guanacaste and across the country. Most of the nonformal norms are encouraged by civil society organizations which have decided to work independently for improving IWRM such as the Fundación Nicoyagua (Fig. 8.4), Confraternidad Guanacasteca, Instituto de Oceanología, or Restoring our Watershed. In the following paragraphs, we provide a summary of the current state of water governance with emphasis on the main economic sectors, i.e., agriculture, domestic consumption, services, industry, and tourism. We highlight bottlenecks identified in the overall governance system and how to explain to them the current state of water resources, as well as the limited efforts for adapting to drought and water scarcity.

The human water consumption in Costa Rica is a mandate for the Costa Rican Institute of Aqueducts and Sewers (Instituto Costarricense de Acueductos y Alcantarillados, ICAA). However, ICAA delegates the administration of aqueducts to semiprivate companies such as the Empresa de Servicios Públicos de Heredia (ESPH),

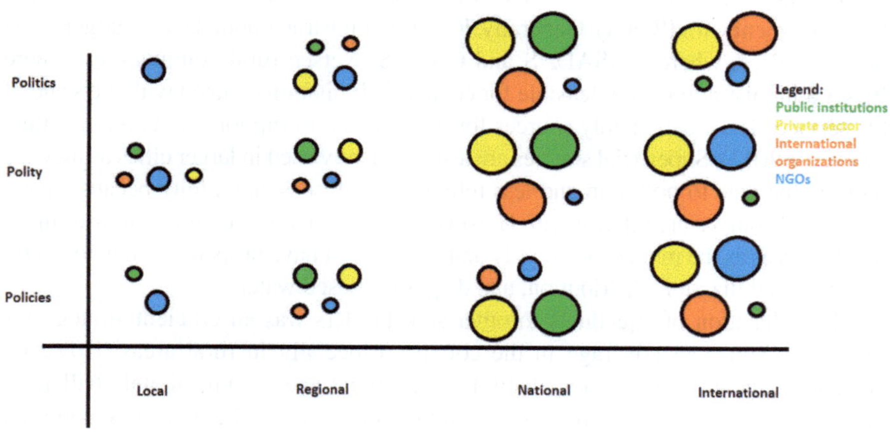

Fig. 8.3 Representation of water governance approaches in Guanacaste and Costa Rica. (Notes: circles represent the leverage of a given sector in a given scale)

Fig. 8.4 Members of the Fundación Nicoyagua and the Commission for the Management of the Potrero and Caimital Basins, Nicoya, Guanacaste, Costa Rica. (Photo: Ricardo Morataya-Montenegro). Note: from left to right: Rolando Castro (CEDARENA), Gerardo Martínez (MINAE), Xinia Campos (MINAE), Gabriela Cuadrado (CEDARENA), Ricardo Morataya (UNA), Juan José Jiménez (private sector representative), Vilmar Rojas (community representative), Nelson Gamboa (Fundación Nicoyagua), Edgar Mora (Fundación Nicoyagua))

to some municipalities, and to Community-Based Drinking Water Associations, locally known as ASADAS or CAARS (less organized committees missing delegation agreement with IICAA). Generally, ICAA administrates aqueducts in larger cities of Costa Rica, whereas ASADAS and CAARS oversee rural communities, where ICAA administration is not feasible for economic limitations related with the scale of the service. The water supply sources for domestic consumption are mainly aquifers (Valverde 2013). Superficial sources are less frequently used in larger cities as they are more vulnerable to pollution and less reliable during seasonal aridity because many superficial sources are intermittent. However, some rural communities still use superficial sources as the main water supply and even do not have proper facilities for water storing, potabilization, distribution, and disposal of used water.

The delegation of aqueducts to other stakeholders was an efficient strategy for increasing aqueduct coverage in the country, especially in rural areas. However, nowadays many delegated aqueducts face technical and organizational challenges that limit their capacity for ensuring that water quality meets the standards requested by the legislation. Moreover, ICAA also lacks the financial and human capacity for fulfilling their regulatory and advisory responsibilities toward delegated systems. Considering the latter, the most recent policy for the administration of rural aqueducts

ICAA proposes the integration of ASADAS and CAARS for reducing their number to facilitate its administration (Astorga et al. 2015). Larger aqueducts in areas with prominent economic development are even targeted for taking over the community-based aqueduct administration. However, the empowerment of the communities with rural aqueducts is that committed that most of ASADAS and CAARS are reluctant to concede their aqueducts to ICAA; thus disputes over the administration of rural aqueducts are generated, e.g., as in Playa Potrero (Navas 2015). Moreover, community-based aqueducts face important constraints for its economical sustainability, as fees are established with a centralized scheme that does not consider scale issues and investment needs required for improving the aqueduct and granting water conservation and sanitation. Beyond water quality and coverage, sanitation is regarded as the most important challenge for IWRM in Costa Rica.

The services, industry, and tourism sectors rely on the water provision of ICAA and on private water concessions, although these sectors also use water provided by delegated systems. Tourism is a business with great economic importance in Guanacaste, higher access to technical advice, technology, and financial resources. Thus, the tourism sector in Guanacaste is searching for options to emancipate from the control of water public institutions, as civil society organizations are questioning the water granted for tourism because the legislation states that human consumption is the top priority. New tourism developments are considering investing in desalinization plants for granting water for tourism activities, even when such technology implies a higher investment and their implementation required the development of national regulations (ICAA 2017; Poder Ejecutivo 2016) and a process for requesting a water concession.

The services, industry, and tourism sectors also rely on private water concessions from superficial sources but mainly from groundwater extracted from artisan and drilled wells. Artisan wells are built without using specialized drilling machines, at a limited depth (< 50 meters detph). The mandate for regulating drilling and underground water concessions relies on the Water Directory (Dirección de Agua, DA), a vice-ministry from MINAE (Ministry of Environment and Energy). Artisan wells require a registration in the DA offices, and water extraction from them does not require a water concession for domestic use or for a limited agricultural use, whereas drilled wells require both a drilling permission and a water concession. Unfortunately, DA has a limited control of the water extractions in legal water concessions. More importantly, DA has limited control on drilling companies and private users for monitoring the construction of unregulated wells. It is estimated that in Guanacaste, there are 10 illegal drilled wells per each well concessioned by DA. Therefore, water balances are always estimating the amount of illegal water extraction, increasing issues for a proper water demand planning. During the months of seasonal aridity, or while being impacted by a hydrological drought, tourism and service sectors may also recur to the purchase of water from delivery companies. In most of the cases, this is a non-regulated business, and there are reports stating that water transportation companies obtain such water without permission and from non-authorized supply sources.

The agricultural sector is the second water user in importance after the energy sector. However, the impact of the agricultural sector in water quality and availability is larger as its use is consumptive. In Guanacaste, large agricultural companies related to the sugarcane production rely on four water supply sources: (i) rainfed; (ii) underground water concessions (i.e., drilled wells); (iii) superficial concessions from the Tempisque River, the largest river in Guanacaste and one of the most overexploited as more than 90% of its flow has been legally granted to the sugarcane production; and (iv) the Arenal-Tempisque Irrigation District (DRAT), which is a 31,000 hectare irrigation project using water imported from an Atlantic watershed. Again, most of the water DRAT is being taken by the sugar and rice agricultural production (Mora Montero 2016), although the most profitable business is aquaculture. Additionally, to sugarcane, rice production is the second crop on extension in the DRAT; however its extension and importance are decreasing as economic integration, liberalization policies, and adaptation to such global changes from the rice and sugarcane industries impose difficulties for rice commercialization (Warner et al. 2018).

In other areas from Guanacaste, melon- and watermelon-exporting farms and livestock farms are among the main water users in Guanacaste. Again, exporting farms and large livestock farmers have a better access and resources for processing water concessions. Therefore, medium and small livestock farmers often rely on artisan wells or unregulated drilled wells, despite two periods of amnesty for encouraging regularization. Moreover, discussing water supply and its regulation with medium and small livestock farmers is always difficult, as they fear facing difficulties while sharing details about their water use. DA has committed investments on decentralization of its services to facilitate the process of drilling permissions and water concessions virtually and using regional offices. However, more efforts are required to facilitate information about such processes, as well as their regulations.

The National Service of Groundwater and Drainage (Servicio Nacional de Aguas Subterráneas y Avenamiento, SENARA) is the administrator of the DRAT. SENARA is organized in four sub-directions: one for research, another for engineering and projects, one more for financial administration, and one for DRAT administration. From all the Costa Rican water institutions, SENARA is the most committed to groundwater research. Additionally, SENARA promotes social processes for improving water governance and using technical information in decision-making by implementing Aquifer Sustainable Use Plans (Planes de Aprovechamiento Sostenible de Acuíferos PAS). However, the governance of the irrigation district has been criticized by the lack of participation of DRAT users (Mora Montero 2016), especially while coping with water scarcity issues (Warner and Kuzdas 2016).

The agricultural sector challenges are related to the implementation of better water management practices, especially for extensive crops such as sugarcane and rice. Moreover, the agricultural sector also must reduce the use of pesticides as Costa Rica has been reported as the major pesticide user per unit of food produced in the world. For instance, sugarcane production is spraying glyphosate at ultralow volume for increasing maturity and sugar concentration. The aerial aspersion of glyphosate on large areas planted with cane represents a high risk for poisoning superficial water sources and nearby towns. In an aspect more related to politics, water concessions and commercial agreements granted for sugar production in Guanacaste are backed up by politicians involved in this business such as the twice president Oscar Arias Sanchez. The Arias Sanchez family is a large investor of one of the most important sugar consortiums (Marchena Sanabria 2016a, b). Moreover, unlike small farmers the agricultural and tourism industries can easily cope with the bureaucratic process required for acquiring a legal water concession and the financial resources needed for its process.

Generally, higher water demands from the domestic and services sectors are rarely questioned. However, all the investment plans announced by tourism, agricultural, and industry sectors are questioned by environmentalist organizations. These actors usually argue that Guanacaste has no water resources for granting the sustainability of economic activities with great water demand or those spatially located in vulnerable regions such as coastal aquifers. The civil society and the government are always confronted while choosing between an apparent dichotomy for favoring economic development and creating employment options or establishing the basis for IWRM.

The amount of technical and scientific information about water has been increasing during the last decade (2009–2019). Particularly, national projects led by Costa Rican public universities, international initiatives such as Futuragua (Belmont forum-funded initiative), and graduate research projects from foreign universities have contributed to improve the understanding of water governance issues. For instance, the doctoral dissertation of Cristopher Kuzdas provided one of the most comprehensive analyses on this matter (Kuzdas and Wiek 2014; Kuzdas et al. 2014, 2015a, b). Kuzdas' work suggests that water governance in Guanacaste can be considered a hierarchical and fragmented hybrid system which struggles as the scale is more decentralized (Kuzdas et al. 2015b). For instance, Kuzdas found a lack of coordination and exchange of information among water stakeholders. Moreover, Kuzdas reported an imbalance in the participation and interest of water stakeholders across water domains. This is because actors are more clustered along domains related with the water use and its delivery, whereas legislation and limited recourses concur to limit the participation on the supply and outflows domains (Kuzdas et al. 2015b).

As we reviewed several challenges remain for improving the water governance system in Guanacaste and Costa Rica. Ironically, such issues are the main factor facilitating changes for accomplishing the latter. For instance, many of the reviewed issues are being addressed with the creation of new legislation and agreements among water stakeholders. Notably, crises have managed to facilitate participatory processes such as the Commission for the Integrated Management from the Nimboyores and Coastal Aquifers from Santa Cruz of Guanacaste (Comisión para el Manejo Integral del Acuífero Nimboyores y Acuíferos Costeros de Santa Cruz de Guanacaste CONIMBOCO). The CONIMBOCO managed to conciliate the different demands of communities, private sectors, ASADAS, and water-related public institutions to engage in a participatory process that allowed an attitude shift from a conflict prone to a scenario where sound agreements are made for building infrastructure and facilitating IWRM. Moreover, despite the repeated failures for enacting a new and progressive water act, other regional and national advances are emerging such as the national decree for water governance (Poder Ejecutivo 2018). This is a presidential decree formalizing a mechanism of participation in water governance. The mechanism establishes three different entities to facilitate the coordination and exchange among water stakeholders: (i) regional forums coordinated with the National Council of Development COREDES (five in total, one per hydrological unit), (ii) a national forum, organized once a year by MINAE, and (iii) the water governance group, as permanent instance that will communicate the recommendation to elaborate water public policy (Poder Ejecutivo 2018). A main issue to be observed is that such instances include the participation and recommendation from underrepresented stakeholders.

8.5 Water-Related Disputes in Guanacaste, Costa Rica

Water scarcity generates scenarios prone to environmental disputes as evidenced in Guanacaste, especially in coastal areas (Esquivel-Hernández et al. 2018; Ramírez Cover 2007). The combination of water scarcity, with a limited water governance, and the power of economic sectors claiming for water among other factors is combined to produce the emergence of socio-environmental conflicts for water resources. Other important factors related to the emergence of water disputes are an unplanned growth of urban centers, the extensive use of agricultural land in rural area, climatic variability, hydrological and geological conditions, the absence of clear and effective policies for the protection of water resources and development planning, and a growing social and environmental inequality that produces inconformity regarding the outcomes of the development models (Ramírez Cover 2007). In this section we present the key learning from recent studies about water-related disputes. For brevity we encourage to review the documentation of water-related disputes in Guanacaste in other specialized works such as Castro Chacón (2004), van Eeghen (2011), Navas (2015), and Navas and Cuvi (2015).

One of the pioneering studies about water-related disputes in Guanacaste identified a spatial and economical pattern (Ramírez Cover 2007). Water-related disputes were more frequently reported in coastal areas, especially those where tourism and the real estate sectors were making large investments. Moreover, this study provides a detailed description of the prominent role of water stakeholders on water disputes. According to Ramírez Cover (2007), the private sector representatives are always accused on judiciary disputes as they produce a shift in the distribution of water resources that may affect nearby communities and ecosystems. However, the same study also highlights normative failures from public institutions that frequently omitted the enforcement of a given legislation. This study also highlights the great contribution from civil society representatives which are always involved in the complaints that originated the disputes.

A follow-up study analyzing five case studies of disputes produced by tourism and agricultural sectors improved our understanding about the conditions that may intensify the severity of a water-related dispute. Specifically, Kuzdas et al. (2016) highlighted the role of leadership, quality of external support, and credibility on water management processes. Kuzdas noticed that efficient leadership will advocate for dispute resolution avoiding its escalation, especially if external stakeholders at a national scale provide support for conflict management. Moreover, credibility becomes an important factor as underrepresented sectors on water governance distrust projects proposed by dominant sectors and water-related public institutions (Kuzdas et al. 2016). The latter will help to explain the successful implementation of CONIMBOCO. More importantly the study suggests that conflict resolution will benefit of a more horizontal governance schemes, especially a polycentric governance system supported by a strengthened human capital among participants. Therefore, improvements in the governance system are expected to reduce inequalities among competing vulnerable sectors and the severity of water-related disputes.

Finally, a third study focused at the national scale provided more insights for understanding water-related disputes. Esquivel-Hernández et al. (2018) showed that in Costa Rica, the main hotspot of water-related disputes is in the Central Valley region, specifically within the boundaries of the Rio Virilla and Rio Grande de Tárcoles watersheds. Moreover, the study showed that many disputes are related to limitations in water infrastructure, especially for water supply and sanitation. Additionally, considering the national scale, water disputes are related to population and the number of hydrometeorological events observed. Surprisingly, wet conditions associated with the wet phase of ENSO (aka La Niña) were the main determinant for water disputes in the Central Valley region (Esquivel-Hernández et al. 2018).

In summary, the development of urban projects in Guanacaste has brought economic benefits such as employment for local inhabitants, but it has also led to considerable environmental damage and friction over the access to strategic resources such as water and land. Action must be taken so that inequality in the access to such resources and the overexploitation of water resources are controlled. Great efforts have been made to address this situation such as the incorporation of the environmental variable in regulatory plans; however, more comprehensive efforts are needed.

References

Astorga Y et al. (2015) Política de organización y fortalecimiento de la gestión comunitaria de los servicios de agua potable y saneamiento. AyA, San José, CR

Castro Chacón S (2004) Del conflicto a la cogestión del agua en la microcuenca del Río Nimboyores, Guancaste, Costa Rica. CATIE

Caviedes CN (1997) Cincuenta años de uso y mal-uso de El Niño Espcacio y Desarrollo:118–135

Chen C-F, Chen C-R, Castellón A, Masferrer M-G-M, Recinos L-E-M (2018) Satellite characterization of drought over cultivated areas in Central America AU - Son, Nguyen-Thanh International Journal of Remote Sensing 39:8505–8526. https://doi.org/10.1080/01431161.2018.1488287

CNE, (Comisión Nacional de Prevención de Riesgos y Atención de Emergencias, CR) (2015) Plan general de la emergencia por sequía según Decreto Ejecutivo N° 38642-MP-MAG, publicado en La Gaceta N° 195, del viernes 10 de octubre del 2014 Enero, 2015. San José, CR
Costa Rica M (2017) Informe de Estado del Ambiente MINAE 713 p
CRRH-SICA (2015) Documentación de la sequía de 2014 y sus impactos en Centroamérica, en los sectores agropecuario, de agua potable y generación hidroeléctrica. San José
Esquivel-Hernández G, Sánchez-Murillo R, Birkel C, Boll J (2018) Climate and water conflicts coevolution from tropical development and hydro-climatic perspectives: a case study of Costa Rica. JAWRA J Am Water Resour Assoc 54:451–470. https://doi.org/10.1111/1752-1688.12617
Hidalgo HG, Amador JA, Alfaro EJ, Quesada B (2013) Hydrological climate change projections for Central America. J Hydrol 495:94–112. https://doi.org/10.1016/j.jhydrol.2013.05.004
Hidalgo HG, Alfaro EJ, Quesada-Montano B (2016) Observed (1970–1999) climate variability in Central America using a high-resolution meteorological dataset with implication to climate change studies. Clim Chang 141:13–28. https://doi.org/10.1007/s10584-016-1786-y
Hidalgo HG, Alfaro EJ, Amador JA, Bastidas Á (2019) Precursors of quasi-decadal dry-spells in the Central America Dry Corridor. Clim Dyn. https://doi.org/10.1007/s00382-019-04638-y
ICAA, (Instituto Costarricense de Acueductos y Alcantarillados) (2017) Especificaciones Técnicas para desalinización y potabilización de agua marina Serie: AyA-2010-01 Alcance 14:27
INEC, (Instituto Nacional de Estadística y Censos, CR) (2012) X Censo Nacional de Población y VI de Vivienda 2011. Características sociales y demográficas vol Tomo I. INEC, San José, CR
Kuzdas C, Wiek A (2014) Governance scenarios for addressing water conflicts and climate change impacts. Environ Sci Pol 42:181–196. https://doi.org/10.1016/j.envsci.2014.06.007
Kuzdas C, Wiek A, Warner B, Vignola R, Morataya R (2014) Sustainability appraisal of water governance regimes: the case of Guanacaste, Costa Rica. Environ Manag 54:205–222. https://doi.org/10.1007/s00267-014-0292-0
Kuzdas C, Warner BP, Wiek A, Vignola R, Yglesias M, Childers DL (2015a) Sustainability assessment of water governance alternatives: the case of Guanacaste Costa Rica. Sustain Sci 11:231–247. https://doi.org/10.1007/s11625-015-0324-6
Kuzdas C, Wiek A, Warner B, Vignola R, Morataya R (2015b) Integrated and participatory analysis of water governance regimes: the case of the Costa Rican dry tropics. World Dev 66:254–268. https://doi.org/10.1016/j.worlddev.2014.08.018
Kuzdas C, Warner B, Wiek A, Yglesias M, Vignola R, Ramírez-Cover A (2016) Identifying the potential of governance regimes to aggravate or mitigate local water conflicts in regions threatened by climate change. Local Environ:1–22. https://doi.org/10.1080/13549839.2015.1129604
Marchena Sanabria J (2016a) Aplicando los credos neoliberales: el impacto de los tratados de libre comercio en el gremio azucarero costarricense, 1994–2010 InterSedes 17:257–290
Marchena Sanabria J (2016b) Las alianzas del azúcar y la política: apuntes para la comprensión histórica de la élite azucarera en Costa Rica (1950-2010). Anu Estud Centroam 42:357–383. https://doi.org/10.15517/aeca.v42i1.26953
Maurer EP, Roby N, Stewart-Frey IT, Bacon CM (2017) Projected twenty-first-century changes in the Central American mid-summer drought using statistically downscaled climate projections. Reg Environ Chang 17:2421–2432. https://doi.org/10.1007/s10113-017-1177-6
Mora Montero JC (2016) Gobernanza del agua en distritos de riego de Costa Rica: El distrito de riego Arenal-Tempisque (DRAT). Cambio, UNA, Universidad Pablo Olavide, Heredia, CR
Morataya-Montenegro R (2011) Dinámica ambiental actual de la Península de Nicoya. Universidad de Costa Rica (UCR)
Navas G (2015) El agua fluye hacia el turista Letras Verdes Revista Latinoamericana de Estudios Socioambientales:27–47
Navas G, Cuvi N (2015) Análisis de un conflicto socioambiental por agua y turismo en Sardinal, Costa Rica Revista de Ciencias Sociales. https://doi.org/10.15517/RCS.V0I150.22835
Obsan-R–SICA (2015) XXVI Foro Regional de Aplicación de los Pronósticos Climáticos a la Seguridad Alimentaria y Nutricional: perspectivas para el período Agosto-octubre 2015
Pahl-Wostl C (2015) Water governance in the face of global change: concepts, methods and practice. Springer, New York. https://doi.org/10.1007/978-3-319-21855-7

Poder Ejecutivo (2016) Reglamento de sistemas de desalinización N° 40098-MINAE-S-TUR Alcance 10:38

Poder Ejecutivo (2018) Crea Comisión para el manejo integrado del Acuífero Nimboyores y Acuíferos Costeros de Santa Cruz de Guanacaste (CONIMBOCO) Alcance

Programa Estado de la Nación (2010) Estado de la Nación 2010

Ramirez P, Bonilla A (2015) Impacto de la sequía en Centroamérica. Contribución especial realizada para el Quinto Informe Estado de la Región

Ramírez Cover A (2007) Conflictos socioambientales y recursos hídricos en Guanacaste UNA descripción desde el cambio en el estilo de desarrollo (1997-2006). Anu Estud Centroam 33/34:359–385

Retana JA, Alvarado L, Araya C, Sanabria N, Solano J, Solera M, Alfaro M (2012) Caracterización del corredor seco en Costa Rica Tópicos Meteorológicos y Oceanográficos 11:18–29

Sánchez-Murillo R, Durán-Quesada AM, Birkel C, Esquivel-Hernández G, Boll J (2017) Tropical precipitation anomalies and d-excess evolution during El Niño 2014-16. Hydrol Process 31:956–967. https://doi.org/10.1002/hyp.11088

Solano J, Villalobos R (w.d.) Regiones y subregiones climáticas de Costa Rica. Instituto Meteorológico Nacional (IMN), San José, CR

UNCCD, FAO, OMM (2013) Política nacional de gestión de sequías. Documento preparado para la Reunión de Al-to Nivel de Políticas Nacionales sobre la Sequía. Ginebra

Valverde R (2013) Disponibilidad, distribución, calidad y perspectivas del agua en Costa Rica. Revista de Ciencias Ambientales 45:5–12

van Eeghen S (2011) Water conflicts in Costa Rica? Sardinal: a case study on the emergence of a water conflict in the context of high speed growth in (residential) tourism. Universiteit Ultrecht

Warner BP, Kuzdas C (2016) Manufactured global-change risk pathways in industrial-based agrarian development. Climate Dev 8:385–396. https://doi.org/10.1080/17565529.2015.1085359

Warner BP, Childers DL, Kuzdas C, Stocks G (2018) Smallholder adaptation to drought in Costa Rica's crony capitalist rice economy development and change. Dev Chang 49:1392–1421. https://doi.org/10.1111/dech.12420

Chapter 9
Integrated Water Resources Management in Iran

Erfan Goharian and Mohamad Azizipour

Abstract Iran is dealing with various water resources challenges. Drying lakes and rivers, declining groundwater resources, water supply rationing and disruptions, agricultural losses, and ecosystem damages are just few of the challenges. This chapter introduces how current management of water resources in Iran led to water crisis in the country, which was formerly renowned as a pioneer of sustainable water management. Iran is located in an arid and semi-arid region, and the combined actions of natural and human factors caused the modern water-related crisis. Although there is no control over natural factors, sustainable water resources planning and management could be achieved by implementing integrated water resources management strategies. This chapter particularly focuses on two of the most vital challenges the country is dealing with: one is mismanagement of the Zayandehrud River basin and the other one is the tragic drying of Lake Urmia. After reviewing the causing problems for each case, we briefly introduce some of the opportunities offered by IWRM practices and identify possible main strategies for the future perspective of management of water resources in Iran.

Keywords Zayandehrud River basin · Lake Urmia · Water resources · Impact

---

E. Goharian (✉)
Department of Civil and Environmental Engineering, University of South Carolina, Columbia, SC, USA
e-mail: goharian@cec.sc.edu

M. Azizipour
Department of Civil Engineering, Faculty of Engineering, Shahid Chamran University of Ahvaz, Ahvaz, Iran

School of Civil Engineering, Iran University of Science and Technology, Tehran, Iran
e-mail: azizipour@scu.ac.ir

E. O. Vieira et al. (eds.), *Integrated Water Resource Management*,
https://doi.org/10.1007/978-3-030-16565-9_9

9.1 Introduction

Iran is located in the Middle East and with the population of about 81 million people and area of about 1,648,195 km^2 is the 18th and 17th most populated and largest country, respectively, in the world (www.amar.org.ir). About 25% of the country is desert and the rest covered by mountains and highlands. The country has a great climate variability. Precipitation ranges from 50 mm to 1000 mm, and temperature ranges from −20 to +50 °C during the year. Iran receives about 250 mm of average annual precipitin, which is about 30% less than the global average. During the past century, this spatial and temporal precipitation variability in Iran has pushed managers toward infrastructure solutions by building dams, large reservoirs, and water diversions and structures to regulate and manage water. There is no doubt that Persians were among the first civilizations and pioneers in managing water resources. Inventing qanat and building water transfer channels and diversions are among the early examples of water management in this region. However, in recent, Iran is dealing with different kinds of water resources problems, including but not limited to drying lakes, fast-paced increasing water demand, decreasing groundwater storage, rationing water supply, degradation of water quality, land subsidence, sandstorms, and ecosystem and environment damages (Madani 2014). Three main drives of these issues are (1) population growth and distribution, (2) inefficient agriculture, and (3) mismanagement and thirst for development (Madani 2014). While these are not the only reasons for recent unsustainable management of the Iran's water resources, individual cases should be solely studied in detail to find specific solution for each of them. There is no silver bullet for the sustainable management of water resources in Iran. In this chapter we review the challenges, current management condition, and future perspective and integrated water resources management opportunities for two important water resources cases in Iran, Zayandehrud River basin and Lake Urmia. Zayandehrud River, which is known as the backbone of human development in central Iran, is drying up and experiencing no water flow. This situation puts into danger the future developments of agriculture, industries, and urban growth in the region. Lake Urmia, which is one of the largest hypersaline in world, has significantly dried due to changes in climate and overallocation of water and excessive water use in upstream (Fathian et al. 2015).

9.2 Zayandehrud River Basin

9.2.1 Description About the Basin

Gavkhuni basin is a part of the Central plateau basin in Iran with the total area of about 41,524 km^2 (Madani and Mariño 2009). The basin is located between the geographical coordinates of 50°-02′ to 53°-22′ east longitude and 31°-12′ to

33°-42′ north latitude. Out of this area, about 40% is mountainous, 59% is lowland, and less than 1% is the Gavkhuni wetland. The Gavkhuni wetland has been recognized internationally under the Convention on Wetlands (1971). However, due to higher water consumption per capita and water demand caused by population growth and industrial development, the wetland does not receive enough water from upstream, and it is shrinking. The minimum flow requirement of Gavkhuni wetland is about 142 MCM/year during normal periods and 60 MCM/year during drought periods (Sarhadi and Soltani 2013). The Zayandehrud basin, with about area, includes six sub-basins upstream of the Zayandehrud Dam, Chelgerd-Ghaleshahrokh (CHGH), Buin-Miandasht (BM), Damaneh-Daran (DAD), Chel-Khaneh (CHKH), Chadegan (CHD), and Yan Cheshmeh (YCH), and ten sub-basins in downstream of the dam, Karvan (kV), Ben-Saman (BS), Alavijeh-Dehagh (ALD), Meimeh (MEIM), Murcheh Khvort (MUKH), North Mahyar (NMHA), Najafabad (NJ), Lenjanat (LJ), Esfahan-Borkhar, and Kuhpaye-Sagzi (KS) (Safavi et al. 2016). Figure 9.1 shows all the sub-basins and the location of Zayandehrud Dam. The hydroclimatic and population distribution in the Zayandehrud River basin is very uneven, for example, the precipitation (annual average precipitation ranges between 500 and 1500 mm), temperature (3–30 °C), and elevation (1470–3974 m a.s.l) vary significantly across the watershed. The basin is surrounded by the mountains and other basins. On the north there is the Salt Lake basin, on east there are the Daghsorkh basin and Kavirsiah Mountain, on south there is the Abarghoosyrjan basin, and the Karoon basin is located on west and southwest of Gavkhuni basin. Zayandehrud River is the main river in Gavkhuni basin and starts in the Zagros Mountains in the southwest of the country. This river flows through the basin and is the main source of water for agricultural, urban, and industrial users' water demand. The river ends in the Gavkhuni swamp, while the remainder of inflow water to the swamp has a decreasing trend over the past few decades.

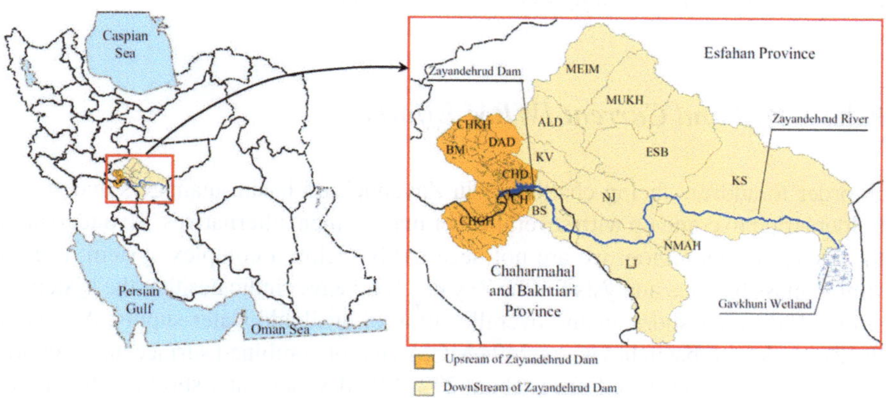

Fig. 9.1 Location of the Zayandehrud basin in Iran and its sub-basins (Safavi et al. 2016)

9.2.2 Facing Problem

The complexity of the Zayandehrud basin system brought the attention of many scientists and researchers to this basin. Different studies tried to understand the glitches in Zayandehrud basin system and provide further insight and explanations why this basin is facing serious water resources issues (Madani and Mariño 2009; Safavi and Bahreini 2009; Safavi et al. 2016). The main issue these days is not having water in the main river of this basin, i.e., Zayandehrud River. However, this is just a symptom of a bigger problem which is the malfunction of the water resources management in this region. To solve this problem, manager and stakeholders for the past few decades looked for traditional- and infrastructure-based solution. For example, conjunctive use of surface and groundwater system in Zayandehrud basin has been one of the key practices to meet increasing demand from different municipal, industrial, agricultural, and environmental sectors in the basin. However, uneven distribution of hydroclimate, demands, and population over the basin and transferring water from Zayandehrud River to other basins exacerbate the problem. Inefficient irrigation system caused high water use in agricultural systems. Due to industrial development and more job opportunities, compared with the neighboring provinces, and readiness of land and water, this region faces growing population and thus higher needs for food, energy, and water. In addition, frequent incidents of droughts in the Zayandehrud River basin have made water resources management a critical issue in the region (Madani and Mariño 2009). Perhaps the transboundary conflict caused by sharing the river by two different provinces can be assumed as the main sociopolitical issue in this region. These two provinces (Esfahan and Chaharmahal and Bakhtiari) are culturally diverse and follow different political and socioeconomic roles (Safavi et al. 2016). The basin has been under intense sociopolitical arguments and faced technical difficulties over the interbasin water transfer projects. Therefore, the basin experiences both engineering and managerial problems along with unpleasant hydroclimate condition.

9.2.3 Past and Current IWRM Efforts

In order to address facing challenges in Zayandehrud basin, managers and stakeholders tried to come up with several water management alternatives. To understand the issues in this region, we are not necessarily facing a complex system; even a simple mass balance analysis for the Zayandehrud River indicates that the system is stressed by water shortage and overallocation of available water supply. Although the Zayandehrud basin has about 2494 MCM/year of combined surface and groundwater resources, the system faces about 272 (MCM-year) water shortage due to the high water demand (total net demand (demand-return flow) of 2766 MCM-year) in the system (Safavi et al. 2016). Water supply for about one-third of the total urban demand is provided by the groundwater system and wells. Out of 16 sub-basins

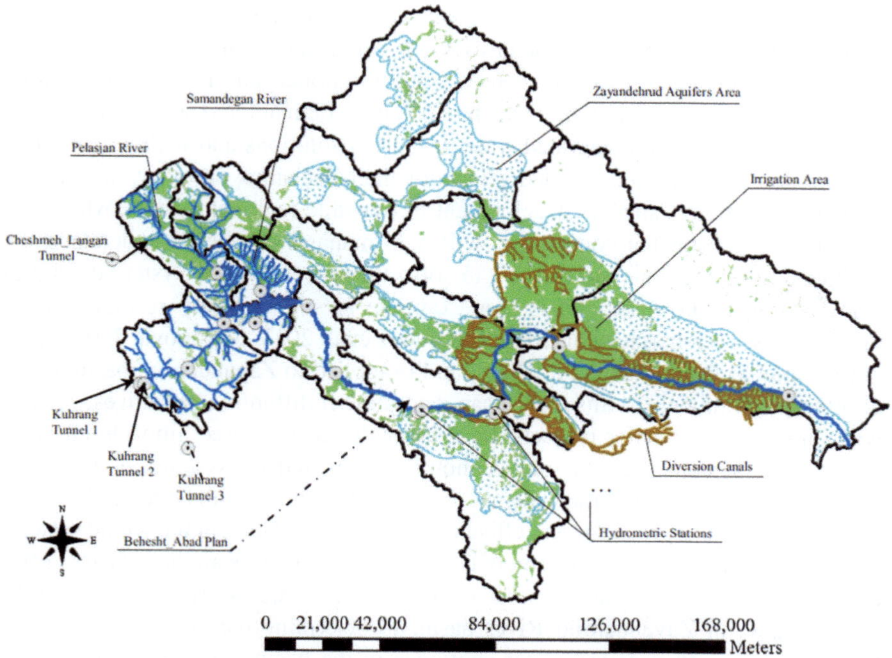

Fig. 9.2 Zayandehrud aquifers and main rivers, irrigation area, diversion canals, and water transfer schemes (Safavi et al. 2016)

(Fig. 9.1), 13 of them are connected to the Zayandehrud aquifer system (Fig. 9.2). To further supply the agricultural and industrial demands, the groundwater system of the Zayandehrud basin is overallocated and leads to overdrafts and decreasing the groundwater level table in this region. Overdrafting groundwater system puts the system in danger for future possible land subsidence, saline water intrusion, loss of natural surface stream, and unreliable water supply for the Zayandehrud basin.

To supply the growing demand in the basin, in addition to mining fossil groundwater resources, interbasin water transfer has been considered as the main management policy. Among the water management alternatives and solutions, water transfer projects have been extensively implemented in respond to increasing water demand in Zayandehrud basin. Currently, three main interbasin tunnels divert water from Karun and Dez basins to Zayandehrud (Gohari et al. 2013). There are two more water transfer projects, Kuhrang Tunnel No. 3 and Beheshtabad, which are untapped due to the arguments over their efficiencies and conflicts between different involved provinces (Fig. 9.2). The main problem about the water transfer, besides the political complexities and difficulties, is that after each new water transfer project, the system starts to build up shortages again (Murray-Rust et al. 2000). These projects impact on the social system and increase their expectations and therefore in a short amount of time increase again the water demand.

Looking at the Zayandehrud basin, we see a great example of a mismanaged system where water managers tried to solve a serious and complex problem, which needs a deep understating of the causes, with traditional- and infrastructure-based solutions. For example, in a basin with unbalanced demand and supply, they try to match the budget by unintegrated water resources solutions and without sufficient information about the unintended effects of their decisions on other parts of the system. Researchers suggest development of dynamic regional models, which can include and present the complexity, will help managers to better understand the problems in this region, and then try to solve them accordingly. System dynamics (SD) approach (Forrester 1997), for example, is suggested by researchers for different cases (Goharian et al. 2016; Karamouz et al. 2013; Simonovic 2003; Stave 2003; Xu et al. 2002; Yang et al. 2008) and is applied to Zayandehrud basin too. A developed SD model for this region can determine if different alternatives, such as water import and transfer projects, have other direct or indirect impacts on other parts of the system such as socioeconomic sector and ecosystems of the river. Integrated management of the water resources system along with better understanding of the causes of problems and effects of decisions will benefit the whole system. Development and application of models for which encompass all these factors offer comprehensive and integrated assessment of complex systems (Tidwell et al. 2004). For example, in Zayandehrud River basin, to determine the efficiency of water transfer projects to decrease water shortages in the basin, an integrated study is required to present physical, social, economic, and political aspects of it. Madani and Mariño (2009), using SD modeling framework, developed a model which shows the dynamic and interrelated characteristics of the system. They used it to first better understand and represent the problem by developing causal loop diagrams (CLDs) of the problem. Then, direct and indirect effects and behavior patterns of the management scenarios have been investigated. Later in this chapter, we will look at different effects of management scenarios and present a future possible perspective of water management for Zayandehrud basin.

9.2.4 Our Conclusion and How IWRM Can Help the Basin

As we discussed before, in quest for employment and more land and water, people from neighboring regions have moved to the Zayandehrud basin. The motivation for further immigration in this basin led to higher water and food demand and started a battle between urban, industrial, and agricultural sectors over water. A crucial fact here is that although managers are trying to provide more water supply, mainly by overdrafting groundwater storage and transfer water, research on these projects shows that after developing each new water transfer project, the system build up shortages again (Murray-Rust et al. 2002). Moreover, water transfer projects affect other sectors in the region. For example, by bringing more water and promoting development in the region, this affects the social system and increases expectations. Ultimately, over time the water demand as well as per capita demand increases and builds up water shortage. Thus, water transfer projects are not sustainable solution to the problems we have.

Clearly, water resource managers need to recognize the interactions between drivers of the problems. If they are informed with better and more reliable knowledge, then they can better understand to what extent trans-basin water diversion can provide benefit to the system without overemphasizing on its application. System dynamics purpose is to represent how and why the dynamics of concern are generated, and then it can evaluate and rank managerial policies (Saysel et al. 2002). Therefore, the first step for managers in this region is to start using systems thinking and better understand what are the actual cause and drivers of the problem.

Looking at the business as usual practices in this region shows that supplying more water from Zayandehrud River basin will lead to an increase in water demand (Gohari et al. 2013). If managers undermine the importance of interrelationships and dynamics of the sub-systems, including hydrologic, socioeconomic, and agricultural sub-systems, then the supply-oriented management alternatives will not suggest sustainable solutions in this basin. Taking into account the possible climate change scenarios, it puts Gavkhuni into a serious danger of not receiving water, and the basin will suffer from extreme water tension (Madani and Mariño 2009). While supply-side management projects are necessary, due to the fact that the system is under a serious water scarcity condition, we anticipate integrated water resources management solutions, which combine the supply and demand management and consider the effects of the policies on other part of system, would draw a sustainable future perspective for Zayandehrud basin. However, unfortunately, managers and planners still are underming this fact and are supplying more water and residing more people in this basin without paying attention to the long-term and unintended consequences of their decisions on ecosystem and environment of the Gavkhuni and Zayandehrud basin.

9.3 Lake Urmia

Lake Urmia is the second largest salt lake in the world with the approximate area of 5000–6000 km^2, located in northwest of Iran as illustrated in Fig. 9.3. The total catchment area of the lake is about 52,000 km^2 including 21 permanent and ephemeral streams.

The lake is mostly fed by rivers, precipitation, and underground springs. Annual inflows from river and precipitation of the lake are 500×10^6 and 1500×10^6, respectively (Jalali 1984). However, the volume of inflow from underground springs is unknown. The main river of catchment is Zarrinehrood with the annual discharge of 2×10^9 m^3.

Lake Urmia basin has a continental climate and is mainly affected by surrounding mountains (Ghaheri et al. 1999; Kelts and Shahrabi 1986). The annual precipitation of the basin varies from 200 to 300 mm with a broad range of air temperature fluctuation from −20° Celsius in winter to 40° Celsius in summer.

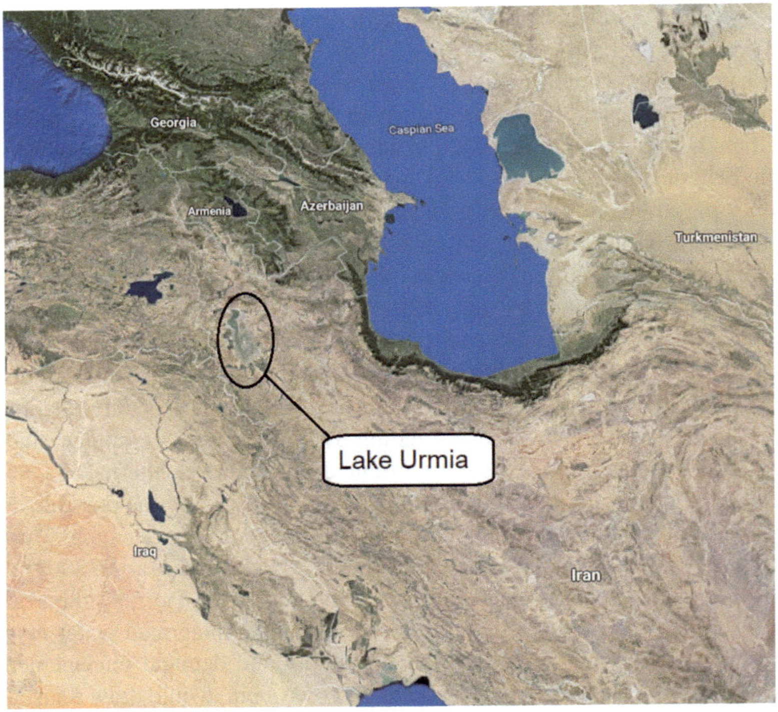

Fig. 9.3 Lake Urmia location in Iran and region

The lake is considered as one of the largest natural habitats of *Artemia* and home to a unique *Artemia* species (Eimanifar and Mohebbi 2007). Brine shrimp is the dominant macro-zooplankton present in many hypersaline environments (Wurtsbaugh and Gliwicz 2001). Günther (1899) described the brine shrimp for Lake Urmia as a unique bisexual species. The ecosystem of the lake has influenced by man-made changes along with natural conditions such that the current status of Lake Urmia is catastrophic.

9.3.1 Anthropogenic Perturbation

During the last decade, the population in the basin increased by approximately 12%. The population density in the basin is about two times greater than average in the country. As a direct impact of population growth, urban water demand is increasing. The population growth in the basin has indirect impacts on water crisis in the area which are more crucial than direct impact. Similarly, food demand and the number of job seekers are increasing. To meet the higher rate of food demand in this region, the cultivated area has been grown up to 40 percent. Since the irrigation system of

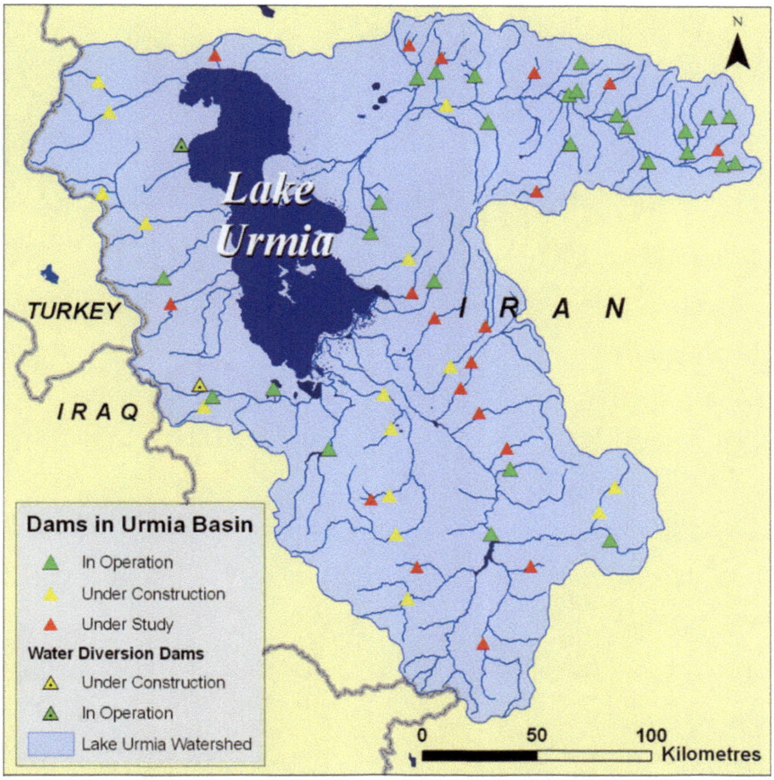

Fig. 9.4 Location of construction projects in Lake Urmia basin

these agricultural lands is inefficient, there has been a huge rise of freshwater demand for irrigation.

As a response to rising water demand, many water structures have been built one after another across the contributory rivers to the Lake Urmia, to store water in reservoirs. Thus, surface water flow to the lake is considerably decreased. Figure 9.4 shows the most important construction projects in the basin. As shown in the figure, unfortunately, there are still numerous either understudy or under-construction structural projects in the basin. These projects are one of the main drivers of decreasing water inflow to the lake which has caused significant problems in the area.

Surface water resources are not the only victims of ever-increasing water demand in the basin. By developing the cultivated area in the basin and due to scarce of fresh surface water resources, groundwater overdraft has occurred during past years. Beyond the other impacts of groundwater overdraft, this also led to decrease inflow to the Lake Urmia.

One of the main projects which have had a drastic impact of the lake's ecosystem is construction of a 15-km causeway to facilitate the regional transportation between Urmia and Tabriz cities. Having only a small gap of 1.2 km, the lake is almost

Fig. 9.5 Lake Urmia and its dividing causeway

divided into two separate parts. Figure 9.5 illustrates the location of the causeway. Since the lake is mostly fed by fresh water from southern part and the northern part has a higher evaporation rate, the causeway has spatially and temporally affected the salinity distribution through the lake. The ecosystem of the lake has disrupted because of changing salinity regime in the lake.

9.3.2 *Facing Problems*

AghaKouchak et al. (2015) used composite multispectral high-resolution (30 m) satellite observations to show the significant shrinkage in the lake's surface area. Figure 9.6 shows the tragic changes of lake's area since 1972. They reported that the area and the volume of Lake Urmia have decreased by about 88% and 80%, respectively.

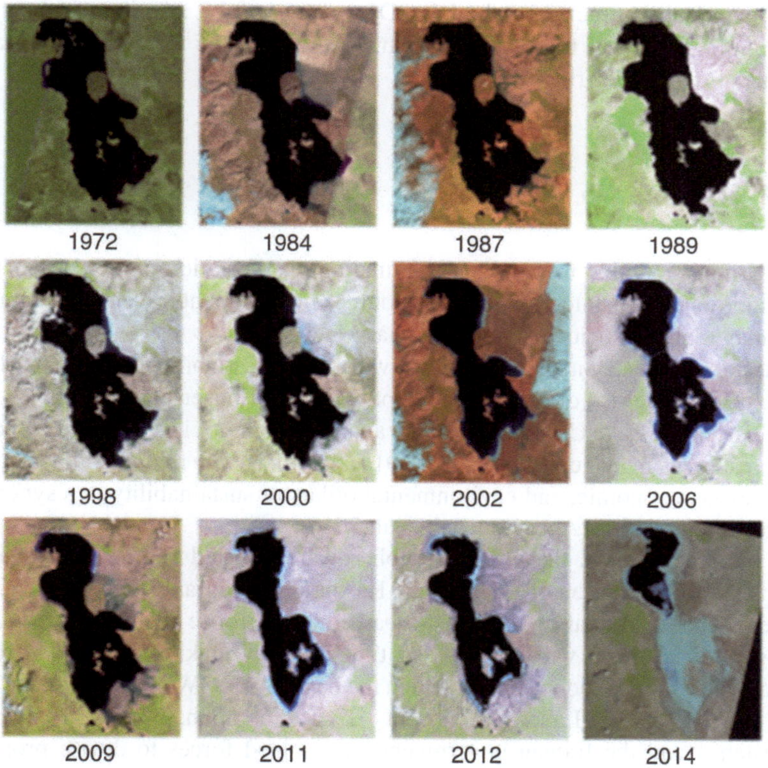

Fig. 9.6 Changes in area of Lake Urmia from 1972 to 2014 (AghaKouchak et al. 2015)

The drying of Lake Urmia has some negative impacts and consequences including agricultural, social, health, environmental, and economic impacts which are mainly driven by salt storms.

The lake's shoreline retreat will increase the frequency of salt storms that sweep across the exposed lakebed, diminishing the fertility of surrounding agricultural lands. The resulting infertile lands will encourage the farmers to move away and cause a lot of social problems.

Effects on humans are perhaps even more complicated. While in the recent past many visitors attracted by the lake believe in its therapeutic properties, the tourism sector has clearly lost out.

Poor air, land, and water quality all have serious chronic health consequences including respiratory and eye diseases. To the best of our knowledge, risks of potential diseases from Lake Urmia salt storms have not been explored at local or regional scales (AghaKouchak et al. 2015).

Salt storms also increased the risk of irreversible ecosystem regime shifts from the lake. Migratory birds such as flamingos, pelicans, ducks, and egrets were historically attracted by the lake. Desiccation of the lake is associated with undermining the local food web by destroying one of the world's largest natural habitats of the brine shrimp *Artemia*.

9.3.3 What to Do

Lake Urmia's looming tragedy is a culminating manifestation of the consequences of uncoordinated, disintegrated water resources management driven and aggravated by managerial and socioeconomic myopia (Madani 2014).

The current management policy along with natural variability and climate change could increase the risk of reaching a "tipping point" (Hansen et al. 2007).

The drying of Lake Urmia can become a turning point for implementing proactive water resources management (Madani 2014) based on deep regional understanding of the social, economic, and environmental pillars of sustainability in a synergistic lake restoration effort (AghaKouchak et al. 2015).

In the recent years, the growing public awareness and calling for restoration actions led to foundation of Urmia Lake Restoration Program (ULRP) by the government. The main goal of ULRP is to pave the way for re-establishing the lake's ecological water level within a 10-year timeframe by making a balance between natural water supply and water demand in the basin. As an IWRM effort, three provinces which share the Lake Urmia basin – East Azerbaijan, West Azerbaijan, and Kurdistan – and the Iranian government have joined forces to devise promising restoration strategies, including stopping new dam construction projects and those in the early construction phase, managing existing reservoirs for the lake restoration only, limiting additional surface water and groundwater withdrawal in the basin, and rent-for-fallowing the surrounding agricultural lands (AghaKouchak et al. 2015). These changes could augment the inflow to the lake and mitigate salt blowouts and sandstorms.

Furthermore, a number of supply-oriented solutions have been proposed, including major water transfers from international transboundary river basins, as well as from the Caspian Sea (UNEP 2012; Zarghami 2011).

Although the ULRP has just started, the bright indications of IWRM, such as increasing lake's water level and birds' immigration, are emerging.

9.4 Future of Integrated Water Resources Management in Iran

Iran is dealing with different kinds of water resources problems. Some of these challenges are not directly related to hydroclimatic conditions and root in mismanagement of resources and sociopolitical and economic situation of the country. As Madani (2014) stated "Iran is experiencing a looming water crisis." Therefore, there is no silver bullet for the sustainable management of water resources in Iran. However, it should not halt the government and managers to emphasize on importance of the issues and recognizes water security as a national priority. Although there are signs of improvement, the motion is still slow. Sustainable management of water resources in Iran requires first better understating of the symptoms instead of looking for immediate solutions. It also requires public awareness and fundamental changes on social behavior and utilization of resources. The complexity of problems and system dictates managers to focus on integrated management of water resources and understand dynamics of coupled human-natural systems. Madani (2014) and Madani et al. (2016) listed main drivers of water crisis in Iran and came up with "Iran's water solutions package." The package includes list of actions and items which are transferable to other cases all around the world. Below is the list of some of these actions:

1. Agricultural modernization.
2. Crop management by considering regional resources availability and economic efficiency.
3. Adjustment of water and energy prices considering socioeconomic impacts and modern technologies.
4. Proactive management of water resources sectors.
5. Emphasizing on education and public awareness.
6. Efficient water market.
7. Incentivizing and forming regional resources management institutions.

These items are still not single solutions to fix water resources issues and to solve water resources problems in Iran; it is necessary to address problems by providing a portfolio which contains concurrent and integrated strategies.

References

AghaKouchak A et al (2015) Aral Sea syndrome desiccates Lake Urmia: call for action. J Great Lakes Res 41:307–311

Eimanifar A, Mohebbi F (2007) Urmia Lake (Northwest Iran): a brief review. Saline Syst 3:5

Fathian F, Morid S, Kahya E (2015) Identification of trends in hydrological and climatic variables in Urmia Lake basin, Iran. Theor Appl Climatol 119:443–464

Forrester JW (1997) Industrial dynamics. J Oper Res Soc 48:1037–1041

Ghaheri M, Baghal-Vayjooee M, Naziri J (1999) Lake Urmia, Iran: a summary review. Int J Salt Lake Res 8:19–22

Gohari A, Eslamian S, Mirchi A, Abedi-Koupaei J, Bavani AM, Madani K (2013) Water transfer as a solution to water shortage: a fix that can backfire. J Hydrol 491:23–39

Goharian E, Burian SJ, Lillywhite J, Hile R (2016) Vulnerability assessment to support integrated water resources management of metropolitan water supply systems. J Water Resour Plan Manag 143:04016080

Günther R (1899) Contributions to the geography of Lake Urmi and its neighbourhood. Geogr J 14:504–523

Hansen J et al (2007) Dangerous human-made interference with climate: a GISS model study. Atmos Chem Phys 7:2287–2312

Jalali B (1984) A report on Urmia Lake Tehran, Ministry of Energy
Karamouz M, Goharian E, Nazif S (2013) Reliability assessment of the water supply systems under uncertain future extreme climate conditions. Earth Interact 17:1–27
Kelts K, Shahrabi M (1986) Holocene sedimentology of hypersaline Lake Urmia, northwestern Iran. Palaeogeogr Palaeoclimatol Palaeoecol 54:105–130
Madani K (2014) Water management in Iran: what is causing the looming crisis? J Environ Stud Sci 4:315–328
Madani K, Mariño MA (2009) System dynamics analysis for managing Iran's Zayandeh-Rud river basin. Water Resour Manag 23:2163–2187
Madani K, AghaKouchak A, Mirchi A (2016) Iran's socio-economic drought: challenges of a water-bankrupt nation. Iran Stud 49:997–1016
Murray-Rust H, Sally H, Salemi H, Mamanpoush A (2000) An overview of the hydrology of the Zayandeh Rud Basin IAERI-IWMI Research Reports 3
Murray-Rust H, Salemi H, Droogers P (2002) Water resources development and water utilization in the Zayandeh Rud basin, Iran IAERI-IWMI Research Report 13
Safavi H, Bahreini G (2009) Conjunctive simulation of surface water and ground water resources under uncertainty Iranian. J Sci Technol 33:79
Safavi HR, Golmohammadi MH, Sandoval-Solis S (2016) Scenario analysis for integrated water resources planning and management under uncertainty in the Zayandehrud river basin. J Hydrol 539:625–639
Sarhadi A, Soltani S (2013) Determination of water requirements of the Gavkhuni wetland, Iran: a hydrological approach. J Arid Environ 98:27–40
Saysel AK, Barlas Y, Yenigün O (2002) Environmental sustainability in an agricultural development project: a system dynamics approach. J Environ Manag 64:247–260
Simonovic SP (2003) Assessment of water resources through system dynamics simulation: from global issues to regional solutions. In: System Sciences, 2003. Proceedings of the 36th Annual Hawaii International Conference on, 2003. IEEE, p 9
Stave KA (2003) A system dynamics model to facilitate public understanding of water management options in Las Vegas, Nevada. J Environ Manag 67:303–313
Tidwell VC, Passell HD, Conrad SH, Thomas RP (2004) System dynamics modeling for community-based water planning: application to the Middle Rio Grande. Aquat Sci 66:357–372. https://doi.org/10.1007/s00027-004-0722-9
UNEP G (2012) The drying of Iran's Lake Urmia and its environmental consequences. J Environ Dev 2:128–137
Wurtsbaugh WA, Gliwicz ZM (2001) Limnological control of brine shrimp population dynamics and cyst production in the Great Salt Lake, Utah. Hydrobiologia 466(1–3):119
Xu Z, Takeuchi K, Ishidaira H, Zhang X (2002) Sustainability analysis for Yellow River water resources using the system dynamics approach. Water Resour Manag 16:239–261
Yang C-C, Chang L-C, Ho C-C (2008) Application of system dynamics with impact analysis to solve the problem of water shortages in Taiwan. Water Resour Manag 22:1561–1577
Zarghami M (2011) Effective watershed management; case study of Urmia Lake, Iran. Lake Reserv Manag 27:87–94

Chapter 10
Water Resources Management in South Korea

Sooyeon Yi and Jaeeung Yi

Abstract This chapter covers the following key topics: an overview of the current state of water resources availability and use characteristics of rivers, large reservoirs, water quality management, water-related natural disasters, and the future water resources management in South Korea. The average annual rainfall in the past 30 years is about 1300 mm, which is greater than world's average annual rainfall, but the spatial and temporal variance is large. Most rivers show characteristics of short lengths and steep slopes, releasing a significant amount of water. These features make the downstream region relatively more vulnerable to massive floods during the wet season. The significant annual fluctuations in water level make water resources development and management difficult. In comparison, South Korea has a larger river regime coefficient than other countries. Therefore, many of these reservoirs are built to store water during the wet season and supply water during the dry season. In the 1960s, South Korea's rapid industrialization has led to a severe deterioration in water quality in most rivers. Since the 1980s, many environmental infrastructures have been built to improve water quality. Therefore, future water resources management strategies in South Korea should focus on (a) establishing a safe and robust foundation for flood control, (b) supplying clean and sufficient water for people and nature, and (c) enhancing sustainable water quality and ecosystem management.

Keywords Water resources availability · River characteristics · Water quality management · Water-related natural disasters · Future water resources management

---

S. Yi (✉)
Department of Landscape Architecture and Environmental Planning,
University of California Berkeley, Berkeley, CA, USA
e-mail: sooyeon@berkeley.edu

J. Yi (✉)
Department of Civil Engineering, Ajou University, Suwon-Si, South Korea
e-mail: jeyi@ajou.ac.kr

E. O. Vieira et al. (eds.), *Integrated Water Resource Management*,
https://doi.org/10.1007/978-3-030-16565-9_10

10.1 Introduction

The Korean Peninsula is located in the Far East. Korea was divided into North Korea (Democratic People's Republic of Korea) and South Korea (Republic of Korea) along the Military Demarcation Line after the Korean War. In 1948, the South Korean government was formally established. South Korea's total area is 100,222 square kilometers, and its total population was 51.6 million in 2016. In Fig. 10.1, the map of East Asia and the Korean Peninsula is presented.

In the past 30 years, the average annual rainfall was 1300 mm, with a broad range from a high of 830 mm to a minimum of 1756 mm. However, the rainfall varies spatially and temporally, which means it fluctuates across the whole country and throughout the entire year. The rainy season typically lasts from June to September, and about 68 percent of total annual rainfall occurs during this season.

Most rivers have short lengths and steep slopes releasing a significant amount of water. These features make the downstream region relatively more vulnerable to massive floods during the wet season. The Miracle on the Han River refers to the period of rapid economic growth in South Korea following the Korean War (1950–1953). During the nation's economic growth, land use transformed from agriculture to other industrial production. Most people who live in areas near the river became very vulnerable to flooding.

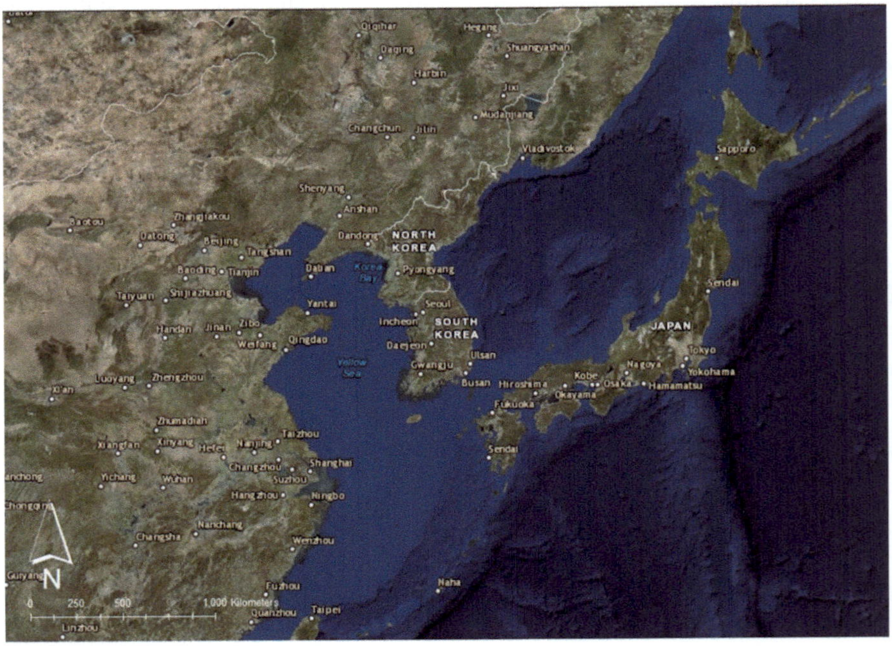

Fig. 10.1 Map of East Asia and the Korean Peninsula

10.2 Water Resources Availability and Current State of Water Use

Over 75.9 billion cubic meters per year (BCM/year) of water is available through the nation. The average water availability is 1488 cubic meters per capita per year. The five major rivers in South Korea are the Han River, the Nakdong River, the Geum River, the Yeongsan River, and the Seomjin River. Out of the five river basins, the average water availability in the Han River and the Nakdong River is relatively low due to the high population density. Figure 10.2 shows water availability in five major rivers.

The yearly average available water is 132.3 BCM. 75.9 BCM (57 percent) is for water use, and the rest is lost through evapotranspiration Fig. 10.3. Out of 75.9 BCM, about 54.8 BCM (72 percent) becomes runoff in the wet season, and 21.2 million cubic meters (MCM) (28 percent) becomes runoff during the rest of the year. The water supply comes from the river (12.2 BCM), the reservoir (20.9 BCM), and the groundwater (4.1 BCM), and the rest flows into the sea (38.8 BCM). Table 10.1 shows the water supply from the reservoir:

The available water is 75.9 BCM per year, of which 4.1 BCM are groundwater supply. The groundwater supply has been increasing, but for the past 4 years,

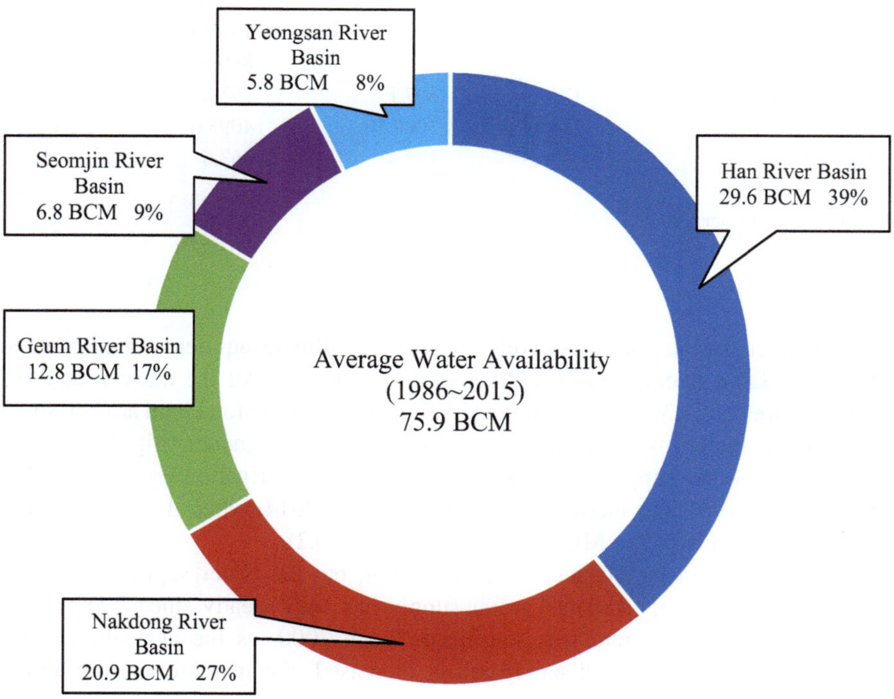

Fig. 10.2 Annual average water availability in five major rivers (MOLIT 2016)

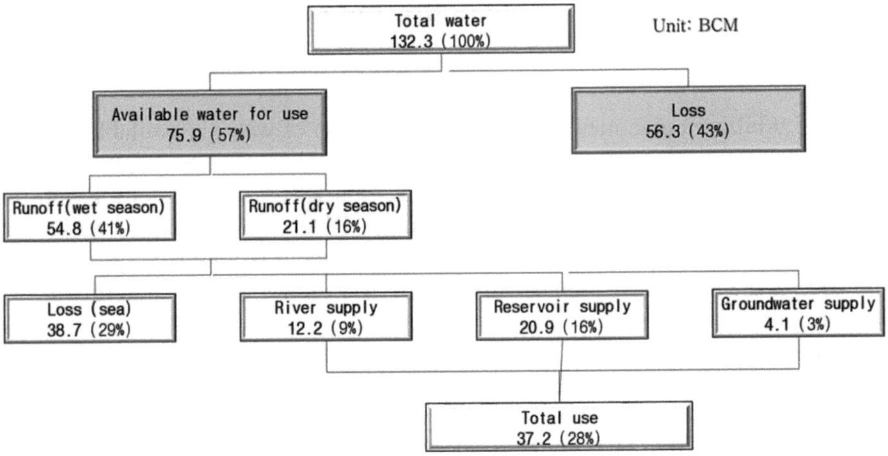

Fig. 10.3 Present state of water use in Korea (MOLIT 2016)

Table 10.1 Main reservoir and weir supply (MOLIT 2016)

	Total storage (MCM)	Effective storage (MCM)	Water supply (MCM/year)	Count
Multipurpose reservoir	12,923.0	9111.0	11,220.2	21
Hydropower reservoir	1844.0	992.8	1335.0	15
Water supply reservoir	609.0	536.3	880.5	54
Estuary bank	1259.3	807.1	2930.0	12
Irrigation reservoir	3142.4	3009.10	4093.0	17,401
Weir (Four Rivers Restoration Project)	626.3	173.4	463.6	16
Flood control reservoir	2709.7	–	–	2
Total	23,113.7	14,629.7	20,922.3	–

groundwater use has been relatively steady. The alluvial aquifers are distributed mostly in large rivers (i.e., Han River, Nakdong River). All the alluvial aquifers cover an area of 27,390 square kilometers (27 percent of the total land area of South Korea). The thickness of aquifers ranges from 2 to 30 centimeters (Fig. 10.4).

The total water supply for domestic, industrial, and irrigation use was roughly 5.1 BCM in 1961 and increased dramatically until 2014, when it was five times greater than in 1961 (251 MCM) (Fig. 10.5, Table 10.2).

Domestic water supply has been increasing, but the water supply per capita has been decreasing. Industrial water supply has been steady since 2000, when the economic growth rate was stabilized. Since 2000, as the agricultural area decreased and the irrigated paddy ration increased, the irrigation water supply started to stabilize.

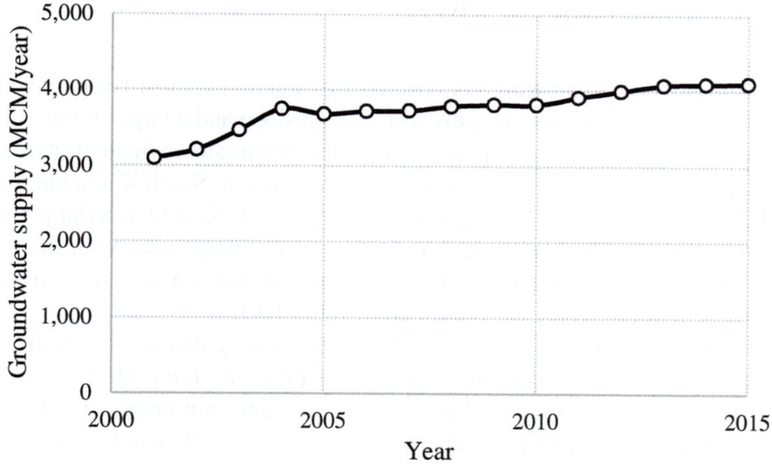

Fig. 10.4 Groundwater supply (unit: MCM/year) (MOLIT 2016)

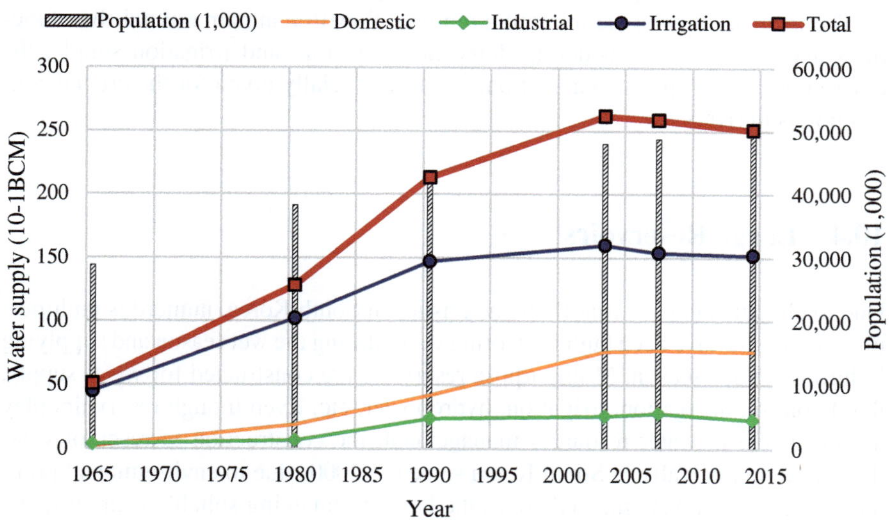

Fig. 10.5 Domestic, industrial, and irrigation water supply (MOLIT 2016)

Table 10.2 Water supply in South Korea (MOLIT 2016)

Water supply (10^{-1} MCM)	1965	1980	1990	2003	2007	2014
Domestic water	2	19	42	76	77	76
Industrial water	4	7	24	26	28	23
Irrigation water	45	102	147	160	154	152
Total	51	128	213	262	259	251
Population (1000)	28,705	38,124	42,869	47,892	48,684	50,747

10.3 Characteristics of Rivers

Since two-thirds of South Korea is covered by mountains, most rivers have small drainage areas and short river lengths with a steep slope and a large amount of sediment. Most major rivers start from the Taebaek Mountains, which stretch across North to South Korea along the Eastern coastline. Eastern South Korea has a steep-sided hill with a short river length, while Western South Korea has a flat plain with a long river length. The Han River has the largest drainage basin with the highest average annual runoff, while the Nakdong River is the longest river (Fig. 10.6). Ten major rivers in South Korea are presented in Table 10.3.

The annual fluctuations in water level are the primary driving forces that make water resources development and management difficult. Table 10.4 is the list of river regime coefficient expressed in the maximum and minimum water flow ratio of major rivers. South Korea has a higher river regime coefficient than many other countries (Table 10.4). Runoff reaches the estuary very rapidly, within 1 to 3 days, right after a massive storm in flood season. Therefore, many reservoirs are built to store this runoff and supply water during the dry season.

The instream flow is the minimum required flow to maintain normal river functions and conditions, considering domestic, industrial, and irrigation supply, the environment, hydropower, navigation, etc. It is officially noted for the crucial control points in rivers.

10.4 Large Reservoirs

Due to climate and river runoff characteristics in South Korea, numerous multipurpose reservoirs have been built for storing water during the wet season and supplying it during the dry season. Multipurpose reservoirs are constructed for water supply, flood control, navigation, irrigation, hydropower, etc. Even though reservoirs play a critical role in water resources management, the construction of reservoirs has decreased dramatically in South Korea since the 2000s due to environmental problems, issues around submerged areas, the lack of remaining suitable locations, etc. In 2016, 20 multipurpose reservoirs were being operated. The Soyanggang reservoir has the largest storage capacity of 2.9 MCM, while the Chungju reservoir has the largest hydropower generation capacity of 410,000 kilowatts. Table 10.5 is a list of major multipurpose reservoirs.

10.5 Water Quality Management

Soon after the industrialization in the 1960s, the urbanization of South Korea progressed rapidly (Table 10.6). The large reservoirs and "wide area water supply systems" are built to satisfy increasing water demands. In South Korea, wide area water

Fig. 10.6 The five large rivers in South Korea

Table 10.3 Ten major rivers and streams in South Korea (WAMIS)

Name	Basin area (km^2)	Length (km)	Average annual rainfall (mm)
Han River	25,954	494	1301
Nakdong River	23,384	510	1186
Geum River	9912	398	1272
Seomjin River	4.96	224	1412
Yeongsan River	3468	137	1318
Anseong Stream	1656	60	1269
Sabgyo Stream	1650	59	1235
Mangeong River	1504	81	1254
Heongsan River	1133	63	1138
Dongjin River	1124	51	1278

Table 10.4 River regime coefficients

River name	River regime coefficient
Han River	90
Nakdong River	260
Geum River	190
Seomjin River	270
Youngsan River	130
Thames River	8
Seine River	34
Rhine River	18
Nile River	30
Mississippi River	3

Table 10.5 Main reservoirs in South Korea (K-water 2016)

River name	Reservoir	Catchment area (km^2)	Height (m)	Length (m)	Total storage (MCM)	Flood control capacity (MCM)	Water supply (MCM)
Han River	Soyanggang	2703	123	530	2900	500	1213
	Chungju	6648	97.5	447	2750	616	3380
	Hoengseong	209	48.5	205	86.9	9.5	119.5
Nakdong River	Andong	1584	83	612	1248	110	926
	Imha	1361	73	515	595	80	591.6
	Hapcheon	925	96	472	790	80	599
	Namgang	2285	34	1126	309.2	269.8	573.3
	Milyang	95.4	89.5	535	73.6	6	73
	Gunwi	87.5	45	390	48.7	3.1	38.3
	Gimcheon-Buhang	82	64	472	54.3	12.3	36.3
	Seongdeok	41.3	58.5	274	27.9	4.2	20.6
	Boheonsan	32.6	58.5	250	22.11	3.49	14.87
Geum River	Daechung	4134	72	495	1490	250	1649
	Yongdam	930	70	498	815	137	650.43
Seomjin River	Seomjingang	763	64	344.2	466	32	350
	Juam (main)	1010	58	330	457	60	270.1
	Juam (auxiliary)	134.6	99.9	562.6	250	20	218.7
Miscellaneous	Buan	59	50	282	50.3	9.3	35.1
	Boryeong	163.6	50	291	116.9	10	106.6
	Jangheung	193	53	403	191	8	127.8

Table 10.6 Urbanization rate (UN 2014)

Year	1960	1970	1980	1990	2000	2010	2050
South Korea (percent)	27.7	40.7	56.7	73.8	79.6	81.9	88.0
World (percent)	32.9	36.0	39.1	43.0	46.6	50.6	66.0

supply systems are defined as a system that delivers raw water or treated water to more than two regions by central or local governments. Also, urbanization caused water contamination and many dry streams.

The Environmental Conservation Act was issued, as the water contamination became a social problem in the early 1970s. By establishing the Environmental Office in 1980, water quality policy was actively enforced. When the Environmental Office expanded to the Environmental Agency after the 1990s, many environmental acts and national plans were established. After the phenol spill happened in the Nakdong River in 1991, the public became highly interested in source water and drinking water safety. In 1994, another water pollution accident occurred in the Nakdong River. This made the water quality policy more regulated by expanding the Environmental Agency to the Ministry of Environment. The total pollution load management system, which restricts the total amount of pollutant load from each basin, was introduced in 1999. The Water Quality Environmental Conservation Act was revised as the Water Quality and Ecosystem Conservation Act in 2005. This includes not only the reduction of water contamination but also the conservation of the water environment and the hydro-ecological system.

By increasing the number of environmental infrastructures, the water quality has been improved, but the enhancement rate has been slowing down recently. The biochemical oxygen demand (BOD) is below 3 mg/l for 95 out of 114 sub-basins in South Korea (MOE 2016). Although the total phosphorus (TP) concentrations have been improving, most regions exceed the Organization for Economic Co-operation and Development (OECD) standard level of TP (0.035 mg/l). The downstream of industrialized areas is still highly contaminated. In 2017, the new administration considers moving the Office of Water Resources from the Ministry of Land, Transport and Maritime Affairs to the Ministry of Environment. One of the reasons is to reinforce the integrated water resources management (IWRM). The Ministry of Land, Infrastructure and Transport was renamed the Ministry of Land, Transport and Maritime Affairs in 2017 when Moon's new administration began.

IWRM is a process that promotes the coordinated development and management of water, land, and related resources to maximize the resultant economic and social welfare in an equitable manner, without compromising the sustainability of the vital ecosystem (GWP 2000). The reasons why IWRM should be introduced in South Korea are as follows: (a) the rainfall varies spatially and temporally, which means that it fluctuates across the whole country and throughout the entire year; (b) the national water management is separated by number of agencies (i.e., the Ministry of Land, Transport and Maritime Affairs; the Ministry of Environment; the Korea Hydro & Nuclear Power Co.; the Ministry of Agriculture, Food and Rural Affairs; and the K-water), which manage the domestic and industrial water supply, water quality management, flood control, hydropower generation, and the irrigation water supply; and (c) laws and systems are unsatisfactory for coordinating the upstream areas with the downstream areas and integrating water quantity and water quality. Implementing IWRM is expected to resolve the problems, for example, the separated water management, water quality and ecosystems in rivers, and unbalanced water supply-demand between regions.

10.6 Water-Related Natural Disasters

Floods and typhoons represent 66 percent of the most severe water-related natural disasters. Flood damage accounted for 59 percent of the water-related natural disaster damage costs. Typhoon damage was the second most costly, accounting for 28 percent. In South Korea, the average water-related disasters cost about 1.8 billion US dollars per year, which is greater than the OECD average. Table 10.7 shows the five worst water-related disasters.

The Imjin River had several major floods (1996, 1998, 2011, and 2013), which caused property damage and human loss. The Imjin River flows from North to South Korea, but the majority of this basin belongs to North Korea (63 percent of the total Imjin River basin). Therefore, South Korea has less control over the Imjin River. North Korea constructed several reservoirs in the upper Imjin River, and they released a huge amount of water downstream several times without warning to South Korea, causing massive damage. The ongoing political tensions between South and North Korea make it difficult to control floods in the Imjin River altogether. The Gunnam flood control reservoir was built in the lower Imjin River to prevent further damage. Even with the construction of the Gunnam flood control reservoir, it is insufficient to control floods in the lower Imjin River because of the short distance between reservoirs in the upstream and downstream, which gives short response time to the downstream reservoir.

In the 2000s, several typhoons struck South Korea, causing serious damage. Typhoon Rusa was the strongest storm on August 23, 2002. A record of 870.5

Table 10.7 Top five heavy rainfalls and typhoons in South Korea (1999–2015)

Rank	1	2	3	4	5
Year	2002	2003	2006	1998	1999
Natural disaster	Typhoon Rusa	Typhoon Maemi	Heavy rainfall and typhoon Ewiniar	Heavy rainfall	Heavy rainfall and typhoon Olga
Date	8/30–9/1	9/12–9/13	7/9–7/29	7/31–8/18	7/23–8/4
Maximum wind velocity (m/s)	Jeju: 43.7 Yeosu: 29.1	Jeju: 60.0	Gunsan: 31.0		Wando: 46.0 Muan: 41.0 Gwangju: 39.6 Masan: 37.0
Maximum daily rainfall (mm)	Gangneung: 870.5 Donghae: 319.5 Sokcho: 295.5	Namhae: 456.3 Goheung: 304	Hongcheon: 255.5 Namhae: 264.5 Sancheong: 229.5	Ganghwa: 481.0 Boeun: 407.5 Yangpyeong: 346.0	Cheorwon: 280.3 Chuncheon: 237.2
Damage cost ($1000)	5,448,148	4,371,790	1,713,418	1,308,717	1,123,817

Rank 1 indicates the most severe event (MOLIT 2015)

millimeters of rainfall in 24 hours was the highest ever recorded since 1904. As a result of Typhoon Rusa, the Janghyeon and Dongmak reservoirs collapsed.

In 2010, frequent localized heavy rainfall inundated the urban areas. Over recent years, urban planning has changed a lot to prepare for urban flooding and to reduce the urban inundation. The design criteria for sewer systems and pump stations were extended from 10-year to 30-year design frequencies. The design frequency for urban rivers was also extended from 50 years to 80 years.

The Korean government amended the Natural Disaster Countermeasures Act to include special disaster areas. The special disaster areas can be declared for all severely harmed areas so that the victims can focus on recovery. The Korean government also revised the River Act to designate the river sections and to introduce emergency action plans for dams.

10.7 Future Water Resources Management

Two-thirds of the total annual rainfall occurs during the rainy season (July–September) and causes floods and droughts, alternately, throughout the year. These climate characteristics make water resources management extremely challenging in South Korea. Therefore, future water resources management strategies should focus on (a) establishing a safe and solid foundation for flood control; (b) supplying clean and sufficient water for people and nature; and (c) enhancing sustainable water quality and ecosystem management.

In order to establish a safe and solid foundation for flood control, it is recommended to analyze climate change impact on floods, plan basin-wide flood reduction strategy, improve the floodplain management plan, and increase the urban flood mitigation capability. In order to supply clean and sufficient water for people and nature, it is recommended to develop a reliable water system through supply and demand management to maintain stable water supply to impoverished water regions and to build a water resources infrastructure based on local characteristics and economic efficiency. In order to enhance sustainable water quality and ecosystem management, it is recommended to establish a national basin-wide environmental management plan, to implement a water quality and environmental monitoring system, and to strictly regulate pollutant sources.

References

GWP (2000) Integrated water resources management, vol 4. Global Water Partnership, SE-105 25 Stockholm, Sweden

K-water (2016) Practical handbook of reservoir operation. K-water, 200, Sintanjin-ro, Daedeok-gu, Daejeon, Republic of Korea

MOE (2016) Environment white book. Ministry of Environment of Korea, Government Complex-Sejong, 11 Doum 6-Ro, Sejong-si, 30103, Republic of Korea

MOLIT (2015) Water disaster mitigation policy and strategy under climate change: case of Korea. Ministry of Land, Infrastructure, Transport of Korea, Government Complex-Sejong, 11 Doum 6-Ro, Sejong-si, 30103, Republic of Korea
MOLIT (2016) Water plan. Ministry of Land, Infrastructure, Transport of Korea, Government Complex-Sejong, 11 Doum 6-Ro, Sejong-si, 30103, Republic of Korea
UN (2014) World urbanization prospects: the 2014 revision-highlights. United Nations, Department of Economic and Social Affairs, Population Division

Chapter 11
Transboundary Groundwater Management and Regulation: Treaty Practices in Africa

María E. Milanés Murcia

Abstract Transboundary groundwater represents an essential source of water for the world population. The management of this precious resource is vital to guarantee the sustainability of regions such as the North-Western Sahara in Africa. International water law instruments such as the 1997 UN Watercourses Convention, the 1992 ECE Water Convention, and the 2008 ILC Draft Articles provide the principles and guidelines to manage transboundary aquifers; however, the type of aquifer determines the legal regimen applicable to it. International groundwater connected to a surface water system is covered by the 1997 UN Watercourses Convention, while fossil aquifers are addressed under the 1994 ILC Resolution on Confined Transboundary Groundwater. Africa is home to some 60 international river basins and over 70 transboundary aquifers. Along the continent, an international watercourse crosses a boundary of every country. Transboundary aquifers represent an important source of water in Africa. Huge reserves of groundwater are located in some of the driest parts of this continent. Many of these watercourses and fossil aquifers are the subject of state practices. Moreover, treaties have been developed between some or all of the riparian states. The trend to regulate transboundary groundwater focuses on agreements addressing mechanisms for exchange of information and scientific research, while the actual management of transboundary aquifers is barely reflected in treaty practices. Only few agreements include in their provisions specific regulations to manage transboundary groundwater in Africa.

Keywords International agreements of water · Management of transboundary groundwater · Transboundary aquifers

M. E. Milanés Murcia (✉)
Sacramento, CA, USA

E. O. Vieira et al. (eds.), *Integrated Water Resource Management*,
https://doi.org/10.1007/978-3-030-16565-9_11

11.1 Introduction

More than half of the world's population depends on groundwater as a primary water source. Irrigation and domestic uses are the main sectors demanding water from aquifers. Increases in population and excessive amounts withdrawn have caused a rapid depletion of groundwater level. Contamination of aquifers is another problem, which makes it very difficult to clean up groundwater; prevention, assessment, and monitoring are essential to protect groundwater resources ensuring the sustainability of this precious element (Moore et al. 1995).

Aquifers and transboundary groundwater have been studied by many different scholars and professionals, but this source remains a mystery for the scientific community. The literature shows a variety of definitions and terms relating to aquifers and groundwater. The interpretation of these definitions and terms creates differing theories and models to manage groundwater. Aquifers can be classified according to the material that forms the structure of the aquifer and their connection to surface water (Moore et al. 1995).

Transboundary groundwater represents an important source in Africa. In fact, Africa is home to over 70 transboundary aquifers (IGRAC 2015). Along the continent, an international watercourse crosses a boundary of every country. Huge reserves of groundwater are located in some of the driest parts of Africa, such as the Nubian Sandstone Aquifer, which is the largest aquifer in the continent (Altchenko and Villholth 2013). The current trend in Africa like in more transboundary aquifers around the world is to provide regulation for the exchange of information of scientific data among countries while the international management and allocation of this valuable resource are still in the process to be adopted by countries (IAEA 2011). The current challenge is to designate and implement "integrated and transboundary water resource management strategies to ensure sustainable access to water for all within Africa" (Semedo 2012).

The special characteristic of groundwater as a hidden source makes it more difficult to regulate and to provide solutions to potential water conflicts among states and countries (IGRAC 2017b). In this regard, the development of international legal instruments regulating transboundary groundwater has been very slow, and it is still in the process to continue the progress of this area of law. This is reflected in treaty practices in Africa and around the world, where only few legal arrangements have been developed reflecting the principle of international water law (Burchi and Mechlem 2005).

The principal international legal instruments addressing the management of transboundary groundwater have been adopted in the recent decades. The 1997 UN Convention on the Law of the Non-navigational Uses of International Watercourses regulates international groundwater connected to a system of surface water (UN 1997). It entered into force on August 17, 2014, and currently 36 countries are party to this Convention (UN Treaty Collection 2017). This Convention codifies the basic principles of international water law: 1) the right, and obligation, to utilize an international watercourse (aquifer) in an equitable and reasonable manner, (2) the duty to take all appropriate measures to prevent the causing of significant harm to other watercourse (aquifer) states, (3) the duty to cooperate, including exchange of data and information, and (4) prior notification of planned measures and consulting and negotiating in good faith. Even if countries are not a party to it, the basic principles of international water law are customary international law and therefore apply to countries (McCaffrey 2006).

Regarding fossil aquifers, the ILC Resolution on confined transboundary groundwater 1994 sets that confined, meaning fossil aquifers, refers to "groundwater not related to an international watercourse" (ILC 1994), which requires precise rules to be applied according to the principles established in the 1997 UN Watercourses Convention (ILC 1994). Fossil aquifers can be covered by the 1992 ECE Water Convention in the case; this type of water resource is intersected by a border (UNECE 1992). This last Convention was opened to global participation on February 6, 2013, and currently 41 countries are party to it (UN Treaty Collection 1992; UNECOSOC 1999). After more than a decade, the International Law Commission adopted the Draft Articles on the Law of Transboundary Aquifers (ILC 2008). When the ILC adopted them, the UN General Assembly (GA) took note of this set of Articles; however, the GA has not determined the eventual application of them. Thus, they remain a draft, which is very similar to the 1997 UN Watercourses Convention, applying its provisions to shared groundwater and providing guidance concerning the management of this resource (ILC 2008; McCaffrey 2010).

The aim of this chapter is to provide an overview of groundwater terminology and show the current legal trend on groundwater management, where Africa is selected as a case study on treaty practices in the regulation and management of transboundary groundwater in the continent. The first part of this chapter contains a selected review of the literature of international transboundary groundwater, including the different terms used to identify aquifer and transboundary groundwater defining different types of aquifers around the world. Then, it addresses transboundary groundwater and its implications in international water law. After that, this chapter focuses on transboundary groundwater resources in Africa addressing the most relevant issues affecting this resource. Then, it provides an analysis of selected treaty practices regarding transboundary groundwater in Africa, showing the current trend in the management of this resource. And finally, the conclusions are presented.

11.2 Defining Aquifers and Transboundary Groundwater

The term aquifer has "different meanings to different people and perhaps different things to the same person at different times" (Freeze and Cherry 1979). It is used to refer to complete geologic formations, to individual geologic layers, and even to groups of geologic formations. Transboundary groundwater has been analyzed in a large number of studies attempting to find the best way to manage this resource. Scholars and professionals from different disciplines have developed a complex and fascinating literature. Hydrologists, geologists, politicians, and lawyers provide different points of view interpreting and defining aquifer and international transboundary groundwater (Eckstein and Eckstein 2004).

According to Moore et al. (1995), an aquifer "is defined as a formation, or part of a formation, containing sufficient saturated permeable material to yield significant quantities of water to wells and springs." These authors also define groundwater as "all subsurface water, as distinct from surface water; that part of the subsurface water in the saturated zone." The real world is complex and has a large variety of hydrological cases that are difficult to classify under one or other type of aquifer.

Freeze and Cherry define an aquifer as "a saturated permeable geologic unit that can transmit significant quantities of water under ordinary hydraulic gradients." These authors also define aquitards and aquicludes. The first "describes the less-permeable beds in a stratigraphic sequence." An aquiclude is "a saturated geologic unit that is incapable of transmitting significant quantities of water under ordinary hydraulic gradients" (Freeze and Cherry 1979).

Moore provided a descriptive classification of aquifers according to the type of material forming the aquifer. This section reproduces below the descriptive view of the most relevant types of aquifers around the world:

1. Geologically recent unconsolidated sand and gravel aquifers are the sources of most of the water pumped in many parts of the world, including North America, the Netherlands, France, Spain and China. Sand and gravel aquifers are common near large to moderately-sized streams; these aquifers were formed by rivers or the meltwater from glaciers.
2. Older sedimentary rocks [which] are usually consolidated by mineralization and the pressure of overlying formations. Sandstone aquifers are formed by the consolidation and cementation of sand. Their porosity ranges from 5 to 30 percent. Their permeability is largely a function of the amount of cement (clay, calcite and quartz). Sandstone aquifers are an important source of groundwater in Libya, Egypt (Nubian Sandstone), the United Kingdom (Permo-Triassic sandstones), north-central United States (St. Peter-Mount Simon Sandstone), and west-central United States (Dakota Sandstone).
3. Limestone aquifers, formed by the consolidation of ocean-bottom calcareous deposits, are the sources of some of the world's largest well and spring yields. Openings that existed when the rocks were formed are frequently enlarged by solution, providing highly permeable flow paths for groundwater. Chalk (a type of limestone) is an important water source in France and the United Kingdom.
4. Basalt and other volcanic rocks also make up some of Earth's most productive aquifers. Basalt aquifers contain water-bearing spaces in the form of shrinkage cracks, joints, and lava caves. Lava tubes are formed when tunneling lava ceases to flow and drain out, leaving a long, cavernous formation. The well yields from volcanic aquifers range from very poor in some regions to some of the most productive aquifers in the world. Recent lavas form the major aquifers in the Hawaiian Islands and the Columbia River Plateau in the northwestern United States.

5. Fractured igneous and metamorphic rock aquifers are the principal sources of groundwater for people living in mountainous areas. Where fractures are numerous and interconnected, rocks can supply water to wells and can be classified as aquifers. Wells are commonly 15–30 meters (50–100 feet) deep. Granite and metamorphic rocks have not been extensively developed as aquifers. Groundwater movement in these rocks is irregular, making exploration for a water supply difficult (Moore et al. 1995).

Aquifers can also be distinguished as confined or unconfined. Confined aquifers are also named artesian aquifers and occur at great depths. This type of aquifer is "contained between two impermeable layers - the base, or "floor," and the "ceiling" strata – that subject the stored water to pressure exceeding atmospheric pressure"(Eckstein and Eckstein 2003). "A confined aquifer is overlain by rocks of lower permeability than the aquifer. The low-permeability layer overlying a confined aquifer is called a confining bed" (Moore et al. 1995). A confining bed has very low permeability; this restricts the movement of groundwater either into or out of the aquifer. In some cases, a spring will result when a fault will allow the passage of water from a confined aquifer to the surface (Moore et al. 1995).

An unconfined area near the ground surface and the water table forms the upper boundary. Another type of unconfined aquifer is a saturated lens bounded by a perched water table (Freeze and Cherry 1979). An unconfined aquifer, also called a water-table aquifer, "is bounded by an impermeable base layer of rock or sediments, and overlain by layers of permeable materials extending from the land surface to the impermeable base of the aquifer"(Eckstein and Eckstein 2003). The recharge to unconfined aquifers is basically by downward seepage through the unsaturated zone. "The water table in an unconfined aquifer rises or declines in response to infiltration of rainfall, pumpage, and changes in stream stage" (Moore et al. 1995). The classification of confined aquifers and unconfined aquifers, related or not to an international watercourse, has large implications in the international legal instrument applicable to the management of transboundary aquifers (ILC 1994), as it is explained in the following section.

11.3 Transboundary Groundwater: Implications in International Law

Groundwater has been included in the definition of "international watercourses" in the 1997 UN Convention on the Law of the Non-navigational Uses of International Watercourses. Article 2(a) defines "watercourse" to mean a system of surface waters and groundwater constituting, by virtue of their physical relationship, a unitary whole and normally flowing into a common terminus (UN 1997). According to (McCaffrey 1991), "international watercourse system" is a term which emphasizes the location of a watercourse system in different states.

> "International watercourse system" […] keeps before the reader the fact that the waters of an international watercourses form a system. This will help to reinforce appreciation of the fact that all components of watercourses are interrelated; and thus, by implication, that it is important to take into account the impact of actions in one watercourse State upon the system-wide condition of the watercourse.

Eckstein (2005) states that the term "system" has not been explicitly defined in the Convention. However, the author assumes that "system" implies an interrelationship between groundwater and surface water where water flows from one to the other resource consistently and in a defined pattern. According to the author, this supposition is supported by the definition of watercourses in the 1997 UN Convention "constituting by virtue of their physical relationship a unitary whole." The author emphasizes that the relationship must also be of a "physical nature" following the ILC's work "that a hydraulic relationship between two surface bodies of water, such as a lake and a connected river, but with no hydraulic connection to any groundwater, also would fulfill the "system" criterion".

The ILC's 1994 Resolution on Confined Transboundary Groundwater considered that groundwater related to an international watercourse was completed on the topic of "the law of the non-navigational uses of international watercourses" and established "the need for continuing efforts to elaborate rules pertaining to confined transboundary groundwater" not related to an international watercourse (ILC 1994). The ILC excludes confined aquifers from the purpose of the 1997 UN Convention. The ILC defined "confined" as groundwater that has no hydrological relationship to surface water (ILC 1994), meaning fossil aquifer. This definition has brought discussion among different disciplines; hydrologists manifest their perspective of the term "confined" as "groundwater relates to groundwater contained and flowing through an aquifer that is under pressure between overlaying and underlain impermeable strata" (Eckstein 2005). "The distinction between confined water, semiconfined water, unconfined water, and perched water is generally a very difficult distinction to make" (Davis and DeWiest 1966). "Groundwater flow is confined when the boundaries or bounding surfaces of the medium (that is, the space made up by the water-filled pores) through which the water percolates are fixed in space for different states of flow" (Davis and DeWiest 1966).

Eckstein also stated that "hydrogeologists know that confined aquifers often are hydraulically connected to and recharged from surface waters in portions of the aquifer that are unconfined, or through lateral flow from higher elevations where the aquifer crops out on the land surface" (Eckstein 2005). According to this last statement "hydraulically connected to and recharged from surface waters" implies connectivity with surface water and, therefore these aquifers would fall into the scope of the 1997 UN Watercourses Convention, which specifically establishes the term system as a physical relationship among surface and groundwater (Krishna and Salman 1999). Moreover, a watershed is composed of surface waters like streams and wetlands and "all the underlying groundwater" (USGS 2016). Unconfined aquifers and the recharge of an aquifer are part of a watershed, which is a surface water system that would make the 1997 UN Watercourses Convention applicable to this type of aquifers. The recharge zone of an aquifer is part of a watershed, which is "an area of land that drains all the streams and rainfall to a common outlet such as the outflow of a reservoir [...] or any point along a stream channel" (USGS 2016). This last definition clearly shows that the recharge zone of an aquifer is part of a surface water system.

Only those fossil aquifers without connection with surface and groundwater are outside of the scope of the 1997 UN Watercourses Convention. Fossil aquifers, also named nonrenewable waters (Krishna and Salman 1999), were formed by recharged water from other geological eras thousands of years ago, and there is no hydrological connection between surface water and other aquifers with the fossil aquifer (Sandoval-Solis et al. 2011).

Nonetheless, the Resolution on Confined Transboundary Groundwater 1994 states "...the principles contained in its draft articles on the law of the non-navigational uses of international watercourses may be applied to transboundary confined groundwater"(ILC 1994). This means that the principles already codified and set on the 1997 UN Watercourses Convention apply to fossil aquifers.

In 2008, the International Law Commission adopted the Draft Articles on the Law of Transboundary Aquifers (ILC 2008). It applies "to the utilization of transboundary aquifers and aquifer systems, other activities that have or are likely to have an impact on those aquifers and aquifer systems; and measures for the protection, preservation and management of those aquifers and aquifer systems" (ILC 2008). The ILC 2008 Draft Article is a useful science-based to manage transboundary aquifer, and in fact the Guarani Aquifer Agreement was developed following the provisions set on it (Guarani Aquifer Agreement 2010). But the scope of the ILC 2008 Draft Articles on the Law of Transboundary Aquifers goes further than the regulation of confined aquifers. The Draft Articles overlap with the 1997 UN Watercourses Convention because both instruments cover groundwater that is connected to a surface water system (McCaffrey 2009).

> The draft, [was] expected to deal with a form of transboundary groundwater not covered by the 1997 U.N. Convention on the Law of the Non-Navigational Uses of International Watercourses, in fact overlaps with the 1997 Convention in terms of the subject matter it covers and thereby gives rise to confusion. (McCaffrey 2009)

The ILC Draft Articles 2008 in Article 3 state "Each aquifer State has sovereignty over the portion of a transboundary aquifer or aquifer system located within its territory. It shall exercise its sovereignty in accordance with international law and the present draft articles" (ILC 2008). The language of Article 3 "refers to a part of an aquifer located within a state" (ILC 2008). The management of the water resources of an aquifer based on the sovereignty concept implies the governance according to the Harmon Doctrine, without considering the special moving quality of water as a fugitive element and therefore the potential harm to other riparian states (Harmon 1895; McCaffrey 1991).

These physical characteristics studied above bring about different political geographical situations, which have been classified into different international groundwater resources law models. The fourth Barberis' models are a legal guideline to identify the transboundary implications of aquifers among countries (Barberis 1986).

1. Confined aquifer without connection with other groundwater or surface water, intersected by international boundary. It is a shore of water alone.
2. Aquifer connected with an international river, but the aquifer is totally in the territory of one state.
3. Two aquifers connected, one is in state A and the other in the neighboring state B (when one of them is connected to a surface water system).
4. The recharge is in state B and the aquifer is located in state A.

The only type of aquifer excluded from the 1997 UN Watercourses Convention is the first model which does not have any connection to a surface water system. The recharge of an aquifer would be evaluated quantitatively according to "(a) precipitation; (b) streamflow-recession displacements; (c) ground-water level fluctuations; (d) age dating of shallow ground water; and (e) watershed characteristics" (Ruhl et al. 2002). The recharge zone of an aquifer is included on a watershed, which is "an area of land that drains all the streams and rainfall to a common outlet such as the outflow of a reservoir […] or any point along a stream channel" (USGS 2016). The watershed is composed of surface waters like streams and wetlands and "all the underlying groundwater" (USGS 2016). As the recharge of an aquifer is part of a watershed, which is a surface water system, it would make the 1997 UN Watercourses Convention applicable to this type of aquifers.

The UNECE Convention on the Protection and Use of Transboundary Watercourses and International Lakes (UNECE 1992) provides special emphasis on environmental protection of groundwater, which is a relevant aspect in ensuring the sustainability of water resources (Art. 3, UNECE 1992). It defines transboundary waters as:

> Any surface or groundwaters which mark, cross or are located on boundaries between two or more States; wherever transboundary waters flow directly into the sea, these transboundary waters end at a straight line across their respective mouths between points on the low-water line of their banks. (Art. 1, UNECE 1992)

The 1992 UNECE Convention sets "guidelines for developing water quality objectives and criteria" establishing "specific requirements regarding sensitive and specially protected waters and their environment." Groundwater is included under the category of "sensitive and specially protected waters" (Annex III, UNECE 1992).

11.4 Transboundary Groundwater in Africa

The International Groundwater Resources Assessment Centre (IGRAC) and the UNESCO International Hydrological Programme have mapped 73 transboundary aquifers in the African continent. Figures 11.1 and 11.2 show the transboundary aquifers identified in Africa in 2015 (IGRAC 2015). IGRAC in collaboration with the British Geological Survey and the University College London developed maps that quantify groundwater resources in Africa. These maps show how groundwater resources differ along Africa (see Fig. 11.3). The development of

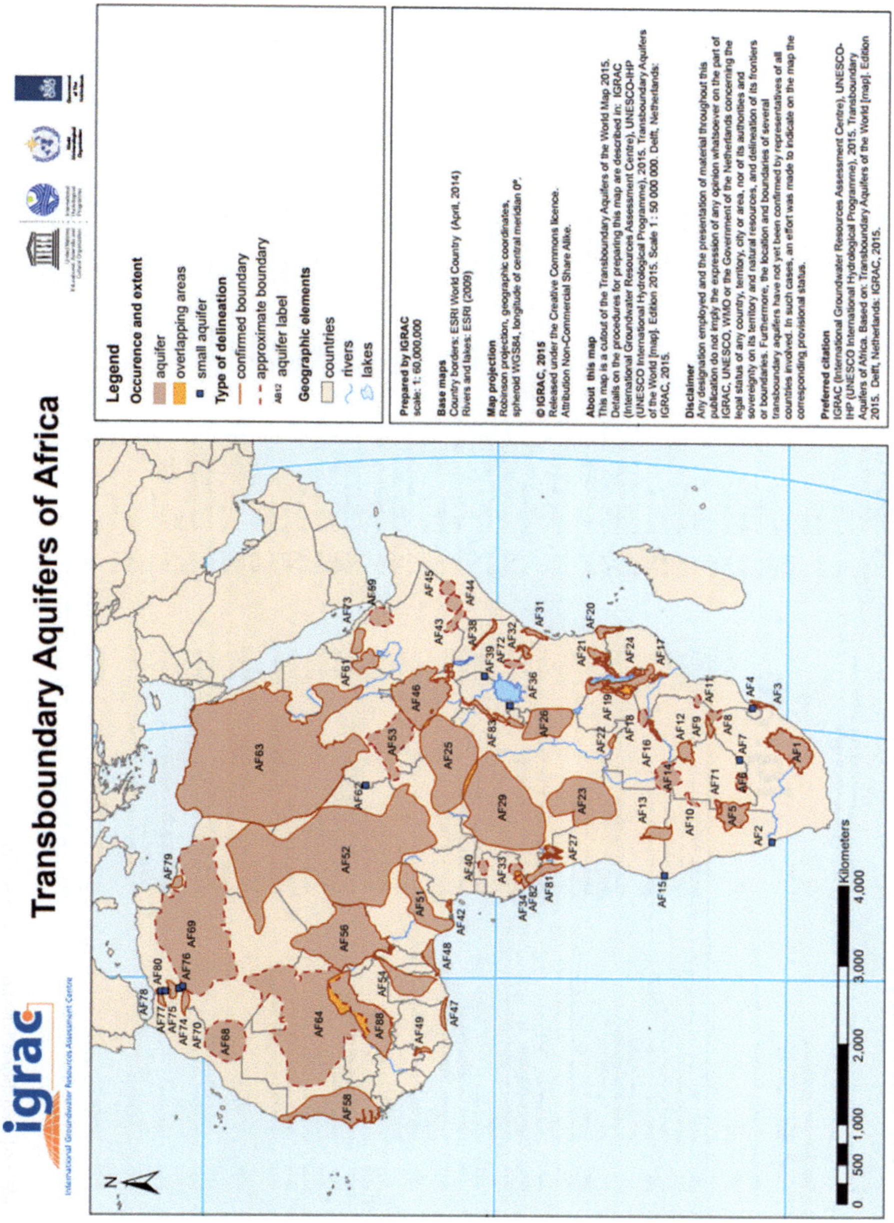

Fig. 11.1 Transboundary aquifers of Africa

Transboundary Aquifers of Africa

For more information on individual transboundary aquifers and the extended view of the small aquifers, please visit IGRAC's Global Groundwater Information System (GGIS) online:

http://tbamap.un-igrac.org/

E-mail: info@un-igrac.org

Westvest 7, 2611 AX Delft, The Netherlands

igrac

Label	Aquifer name	Sharing countries	Area [km²]
AF1	Karoo Sedimentary Aquifer	Lesotho, South Africa	165,936
AF2	Coastal Sedimentary Basin V	South Africa, Namibia	913
AF3	Coastal Sedimentary Basin VI / Coastal Plain Sedimentary Basin Aquifer	Mozambique, South Africa	10,230
AF4	Rhyolite-Breccia Aquifer	South Africa, Swaziland, Mozambique	4,916
AF5	Stampriet Aquifer System	Botswana, Namibia, South Africa	102,401
AF6	Khakhea/Bray Dolomite	Botswana, South Africa	29,689
AF7	Zeerust / Lobatse / Ramotswa Dolomite Basin Aquifer	Botswana, South Africa	351
AF8	Limpopo Basin	Mozambique, South Africa, Zimbabwe	19,961
AF9	Tuli Karoo Sub-Basin	Botswana, South Africa, Zimbabwe	14,330
AF10	Northern Kalahari / Karoo Basin / Eiseb Graben Aquifer	Botswana, Namibia	12,336
AF11	Save Aluvial	Mozambique, Zimbabwe	11,477
AF12	Eastern Kalahari Karoo Basin	Botswana, Zimbabwe	39,603
AF13	Cuvelai and Ethosa Basin / Ohangwena Aquifer System	Angola, Namibia	47,131
AF14	Nata Karoo Sub-basin / Caprivi deep-seated Aquifer	Angola, Botswana, Namibia, Zambia, Zimbabwe	90,982
AF15	Coastal Sedimentary Basin IV	Angola, Namibia	1,352
AF16	Medium Zambesi Aquifer	Zambia, Zimbabwe	10,705
AF17	Shire Valley Alluvial Aquifer	Mozambique, Malawi	6,223
AF18	Arangua Alluvial	Mozambique, Zambia	21,235
AF19	Sand and Gravel Aquifer	Malawi, Zambia	25,318
AF20	Coastal Sedimentary Basin III	Mozambique, Tanzania	25,344
AF21	Karoo Sandstone Aquifer	Mozambique, Tanzania	40,007
AF22	Kalahari/Katangian Basin/Lualaba	Zambia, Democratic Republic of the Congo	7,880
AF23	Coango	Democratic Republic of the Congo, Angola	367,328
AF24	Weathered basement	Tanzania, Zambia,Malawi	126,590
AF25	Karoo-Carbonate	Central African Republic, Congo, South Sudan	604,596
AF26	Tanganyika	Burundi, Democratic Republic of the Congo, Tanzania	184,595
AF27	Dolomitic Basin	Angola, Democratic Republic of the Congo, Congo	21,316
AF29	Cuvette	Congo, Democratic Republic of the Congo	876,603
AF31	Coastal Sedimentary Basin I / Karoo Sedimetary Aquifer	Kenya, Tanzania	17,067
AF32	Kilimanjaro Aquifer	Kenya, Tanzania	14,579
AF33	AF33	Congo, Gabon	23,543
AF34	AF34	Congo, Gabon	7,151
AF36	Kagera Aquifer	Tanzania, Rwanda, Uganda	5,779
AF38	Merti Aquifer	Kenya, Somalia	13,623
AF39	Mount Elgon Aquifer	Uganda, Kenya	5,398
AF40	AF40	Congo, Gabon	19,681
AF42	Rio Del Rey	Nigeria, Cameroon	6,442
AF43	Dawa	Ethiopia, Kenya, Somalia	34,007
AF44	Jubba	Ethiopia, Somalia	34,587
AF45	Shabelle	Ethiopia, Somalia	30,985
AF46	Sudd Basin	Ethiopia, Kenya, South Sudan	370,648
AF47	Tano Basin	Ghana, Cote d'Ivoire	16,063
AF48	Keta / Dahomey / Cotier Basin Aquifer	Ghana, Togo, Benin, Nigeria	36,904
AF49	Cestos - Danané Aquifer	Cote d'Ivoire, Guinea, Liberia	9,403
AF51	Aquifer Vallee de la Benoue	Nigeria, Cameroon	219,001
AF52	Lake Chad Basin	Chad, Niger, Nigeria, Cameroon, Central African republic, Algeria	2,271,303
AF53	Baggara Basin	Central African Republic, South Sudan, Sudan	239,877
AF54	Volta Basin	Benin, Burkina Faso, Ghana, Togo, Niger	144,277
AF56	Irhazer-Illuemeden Basin	Algeria, Benin, Mali, Niger, Nigeria	577,885
AF58	Senegalo-Mauretanian Basin	Gambia, Guinea-Bissau, Senegal, Mauritania, Western Sahara	332,570
AF59	Afar Rift valley / Afar Triangle Aquifer	Djibouti, Ethiopia	57,011
AF61	Gedaref	Ethiopia, Sudan	57,831
AF62	Disa	Chad, Sudan	1,482
AF63	Nubian Sandstone Aquifer System (NSAS)	Chad, Egypt, Libya, Sudan	2,892,867
AF64	Taoudeni Basin	Algeria, Mali, Mauritania	1,260,940
AF68	Système Aquifère de Tindouf	Morocco, Western Sahara, Mauritania, Algeria	221,019
AF69	Northwest Sahara Aquifer System (NWSAS)	Algeria, Libya	1,279,963
AF70	Système Aquifère d'Errachidia	Morocco, Algeria	20,721
AF71	Ncojane Basin	Botswana, Namibia	10,323
AF72	Rift Aquifer	Kenya, Tanzania	21,145
AF73	Mereb	Ethiopia, Eritrea	38,753
AF74	Angad	Morocco, Algeria	4,677
AF75	Ain Beni Mathar	Morocco, Algeria	18,315
AF76	Chott Tigri-Lahouita	Morocco, Algeria	3,560
AF77	Figuig	Morocco, Algeria	1,546
AF78	Jbel El Hamra	Morocco, Algeria	561
AF79	Système Aquifère de la Djeffara	Tunisia, Libya	16,627
AF80	Triffa	Morocco, Algeria	11,530
AF81	Aquifere Cotier	Angola, Democratic Republic of the Congo, Congo, Gabon	45,531
AF82	AF82	Gabon, Congo	19,012
AF83	Aquifere du Rift	Democratic Republic of the Congo, South Sudan, Uganda	44,632
AF88	Aquifer extension Sud-Est de Taoudeni	Mali, Guinea, Burkina Faso	343,482

Fig. 11.2 Code and transboundary aquifers in Africa

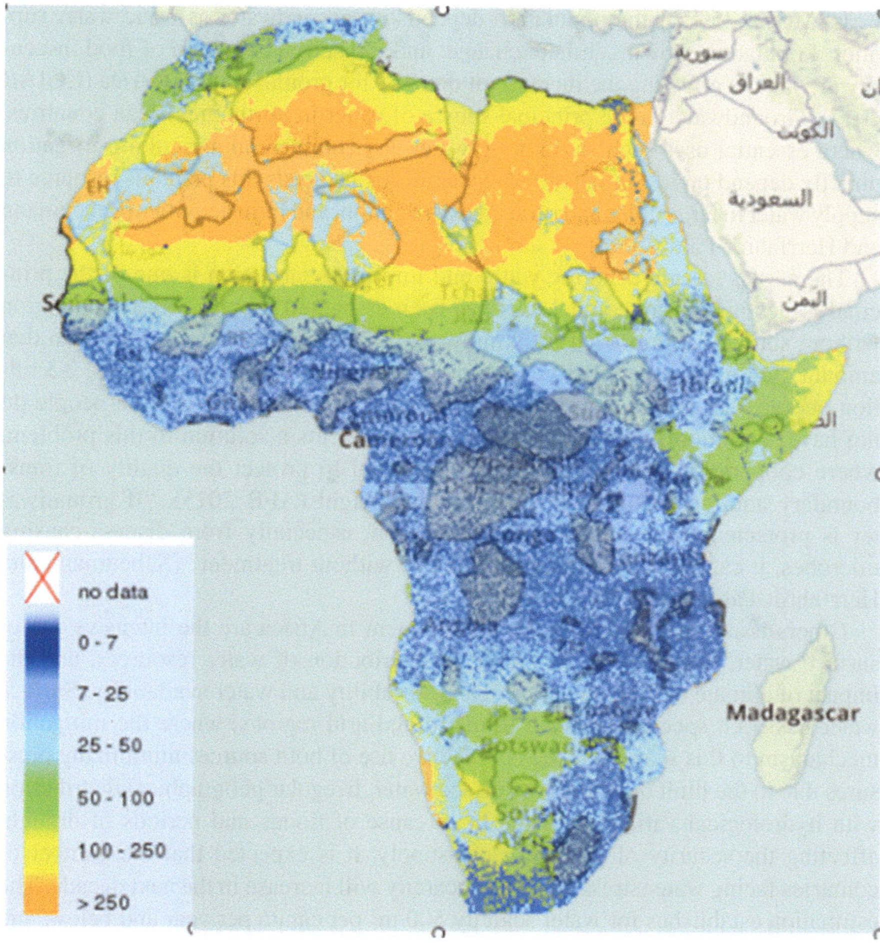

Fig. 11.3 Estimated depth to groundwater (mbgl) and transboundary aquifer of Africa 2015. (Source: https://ggis.un-igrac.org/ggis-viewer/viewer/groundwaterafrica/public/default)

these maps provides information about resilience to climate change and how groundwater can be applied "in adaptation strategies to climate and other environmental changes" (IGRAC 2017a). However, further scientific research is still necessary to determinate aspects such as the impact of runoff on groundwater, the type of aquifer, and the connectivity as part of a surface water system (Camacho Suarez et al. 2015).

Transboundary groundwater in Africa is an essential source for the development of this continent (Altchenko and Villholth 2013). Transboundary aquifers cover approximately 42% of Africa and provide water to 30% of the population (CGIAR 2017). The special and relevant aspect of transboundary groundwater on governance and management has been recently recognized as an essential element to be included in water management (CGIAR 2017).

In Africa, 75% of the population depends on groundwater as basic water supplies. Population growth, climate change, and the current situation of food insecurity are factors affecting the increase of demand for groundwater in Africa (CGIAR 2017). Groundwater is the principal source of water in northern African countries, where essential uses such as drinking water and commercial agriculture irrigation directly depend on it. Similarly, sub-Saharan Africa needs this precious resource to supply water to large cities and rural areas, irrigation, and farm activities (Scheumann and Herrfahrdt-Pähle 2008).

The access to safe drinking water and improved sanitation is one of the principal issues in African countries. It has been estimated that the world population without sources of improved drinking water is 768 million people, and from that amount, 344 million people live in Africa. Regarding sanitation facilities, 2.5 billion people lack access to it in the world. Only in Africa, 612 million people do not have sanitation facilities. Groundwater represents a solution to this problem, where cooperation among countries is essential to protect the quality of transboundary aquifers ensuring this basic human right (AFB 2015). "If groundwater is protected from anthropogenic pollution, especially from disease-causing microbes, it can often be consumed directly without treatment" (Scheumann and Herrfahrdt-Pähle 2008).

Others issues in terms of water management in Africa are the intensive use of surface water, the irregular geographical distribution of water resources, and the impact of climate change on hydrological variability and water availability. Surface water has been specially used in semiarid and arid regions, where the mitigation mechanism to this scarcity is the conjunctive use of both sources minimizing pressures due to the limit quantities of surface water. Irregular geographical distribution with hydrological variability is the main cause of floods and periods of drought affecting the security of reliable water supply. It is expected that the number of countries facing water stress and water scarcity will increase in the next decade. The estimation establishes for water scarcity 500 m^3 per capita per year and below, and for water stress the estimation is below 1700 m^3 per capita per year by 2025 (Scheumann and Herrfahrdt-Pähle 2008).

11.5 The Current Trend on the Management of Transboundary Groundwater Reflected on Treaty Practices in Africa

State practice has reflected the trend toward greater focus on groundwater. Although few agreements include groundwater regulation, the trend demonstrates how recent international treaties focus on groundwater or have specific provisions to manage groundwater (Mechlem 2016). For example, the two agreements on the Nubian Sandstone Aquifer System between Chad, Libya, and Sudan signed in 2000 have as their principal scope the cooperation through data collection and exchange of information for aquifer modeling (FAOLEX 2000). Another example specifically addressing groundwater is the North-Western Sahara Aquifer System (SASS) between Algeria, Libya, and Tunisia (Burchi and Mechlem 2005). The main purpose of this last agreement is to institutionalize a joint commission for data collection and exchange for aquifer modeling. It is the result of long-term cooperation between the three countries to share the aquifer in an equitable and reasonable manner, without causing harm to other states (Burchi and Mechlem 2005). This section shows how the trend is to regulate transboundary groundwater and include the principles of international water law in treaty practices in Africa.

11.5.1 Agreements Between Chad, Egypt, Libya, and Sudan on the Nubian Sandstone Aquifer System (2000)

The Nubian Sandstone Aquifer System extends approximately 2 million square kilometers and is the only source of water for those desert regions far away from the Nile River. It is shared by Chad, Egypt, Libya, and Sudan (IAEA 2011). These countries signed two agreements in 2000 for exchanging data and information to monitor the situation of the aquifer in Tripoli October 5, 2000.

Agreement No. 1: Concerning the monitoring and exchange of information related to the groundwater of the Nubian Sandstone Aquifer System, which established sharing and access to information only for the four NSAS countries.

Agreement No. 2: For monitoring and sharing data for the sustainable development and proper management of the Nubian Sandstone Aquifer System, established continuous yearly monitoring of the aquifer, as well as the taking of electrical conductivity measurements and water level.

The main goal of the efforts of these agreements was data collection and exchange of information for aquifer modeling and establishing joint institutions. These agreements reflect the principle of cooperation, notification of planned measures, and consulting and negotiating in good faith. Within the two agreements, the four countries share data through the implementation of them. In addition, other information such as socioeconomic data, management of a harsh environment, drilling experiences, and meteorological data is shared through an Internet environment – server and access through the Internet. Oracle is the web-based site where data is stored by the four countries. Sustainable development and proper management of the Nubian Sandstone Aquifer System are done through continuous monitoring that is shared among the countries. The monitoring network is subject to changes based on the feedback of the National Coordinators of the concerned countries (FAOLEX 2000). The Nubian Aquifer has been qualified as a confined aquifer (fossil aquifer), which would be within the scope of the 2008 ILC Draft Articles on Transboundary Aquifers (Burchi and Mechlem 2005).

In 2013, the four countries signed the Regional Strategic Action Programme for the Nubian Aquifer System. This agreement enhances cooperation management of the shared aquifer providing strategies involving the Joint Authority. The vision adopted by the four countries is "to assure rational and equitable management of the [aquifer system] for sustainable socio-economic development and the protection of biodiversity and land resources whilst ensuring no detrimental effects on the shared aquifer countries" (IAEA and UNDP–GEF 2013).

11.5.2 "Establishment of a Consultation Mechanism for the North-Western Sahara Aquifer System": Algeria, Libya, Tunisia (2002)

The North-Western Sahara Aquifer System (NWSAS) extends over 1 million square kilometers including the Intercalary Continental and the Terminal Complex Aquifers (International Waters Governance 2017). The agreement between Algiers, Tripoli, and Tunis about Establishment of a Consultation Mechanism for the North-Western Sahara Aquifer System was the result of a meeting held at the Headquarters of the Food and Agriculture Organization (FAO) of the United Nations in Rome, Italy, on December 20, 2002. The minutes and letters of endorsement from each country meant an agreement establishing the Consultation Mechanism (Burchi and Mechlem 2005).

The scope of this agreement is limited to forming a joint institution for data collection and exchange for aquifer modeling. The objective is "to coordinate, promote and facilitate the rational management of the NWSAS water resources." Management and organization are structured by a committee formed of members of the national agencies with authority on water resources. Among the main functions, it is possible to emphasize the development of "databases on socio-economic activities in the region," as well as publishing "indicators on the resources and [their] use in the three countries." The financial coordination is based on cost supported by each state. Cooperation, exchange of information, measures to prevent harm, and equitable and reasonable manners managing water resources are the basic principles of international water law emerging from this agreement (Burchi and Mechlem 2005).

A coordination unit for the Consultation Mechanism for the North-Western Sahara Aquifer System, which was established in 2008, continues updating the database with a total of 16,500 water gates in 2015. Additional efforts have been developed to monitor water table, and an initiative addresses control on the water quality of the aquifer (OSS 2016).

11.5.3 Revised Protocol on Shared Watercourses in the Southern African Development Community (SADC) (2000)

The Protocol is a regional treaty signed by 13 countries in Windhoek in August 7, 2000, and entered into force in September 22, 2003 (ECOLEX 2003b; SADC 2000). The Protocol follows the general principles and guidelines codified on the 1997 UN Convention. "The objective of this Protocol is to foster closer cooperation for judicious, sustainable and co-ordinated management…" (Art. 2). It specifically states that "State Parties undertake to respect the existing rules of customary or general international law relating to the utilization and management of the resources of shared watercourses" (Art. 3.3). Groundwater is included in Art. 1, which uses the term watercourses meaning "a system of surface and groundwater consisting by virtue of their physical relationship a unitary whole normally flowing into a common terminus such as the sea, lake or aquifer." This definition is similar to the definition set on the 1997 UN Watercourses Convention. The Protocol regulates groundwater in conjunctive use with surface water and establishes an institutional framework for the implementation of all principles and practices. The framework includes shared watercourse institutions to contribute to the equitable and reasonable use of water (Burchi and Mechlem 2005; SADC 2000).

11.5.4 Tripartite Interim Agreement Between the Republic of Mozambique, the Republic of South Africa, and the Kingdom of Swaziland for Cooperation on the Protection and Sustainable Utilization of the Water Resources of the Incomati and Maputo Watercourses

The Tripartite Interim Agreement between the Republic of Mozambique, the Republic of South Africa, and the Kingdom of Swaziland for Cooperation on the Protection and Sustainable Utilization of the Water Resources of the Incomati and Maputo Watercourses was signed in Johannesburg, August 29, 2002 (ECOLEX 2002).The Republic of Mozambique ratified it on December 1, 2004 (ECOLEX 2002). The preamble takes into account the principles and norms reflected on the 1997 UN Watercourses Convention. This agreement, like the previous one above, defines "watercourse" similar to the 1997 UN Watercourses Convention, as a system including surface and groundwater constituting a unitary whole normally flowing into a common terminus such as sea, lake, or aquifer. The main goal of this agreement (Art. 2) is cooperation among the Parties to ensure the protection and sustainable utilization of the water resources of the Incomati and Maputo watercourses. Article 13 and Annex II specifically refer to transboundary impacts in aquifers and the limits for recharging and abstraction facilities (ECOLEX 2002).

11.5.5 The Convention on the Sustainable Development of Lake Tanganyika (2003)

The Convention on the Sustainable Development of Lake Tanganyika was signed by Burundi, Democratic Republic of the Congo, Zambia, and the United Republic of Tanzania in Dar es Salaam, June 12, 2003. It entered into force by Burundi and the United Republic of Tanzania in August 23, 2005 (ECOLEX 2003a). The main goal of this convention is "to ensure the protection and conservation of the biological diversity and the sustainable use of the natural resources of Lake Tanganyika and its Basin by the Contracting States of the basis of integrated and co-operative management" (Art. 2) (ECOLEX 2003a). The "Lake Basin" includes the whole "system of surface waters and groundwater that flow into the Lake from the Contracting States and the land submerged by these waters." This convention lays down obligations according to the principles of preventive action, participation, fair and equitable benefit, precautionary principle, and the polluter pays principle (Art. 5) (ECOLEX 2003a). Prior notification, public participation in decisionmaking processes, and exchange of information are also required to be followed by all Contracting States (Burchi and Mechlem 2005).

11.5.6 Protocol for Sustainable Development of Lake Victoria Basin (2003)

The Protocol for Sustainable Development of Lake Victoria Basin was signed by the East African Community, the Republic of Kenya, the Republic of Uganda, and the United Republic of Tanzania in Arusha November 29, 2003 and entered into force November 30, 2004. Burundi and Rwanda ratified it in June 18, 2007 (ECOLEX 2003b). This treaty has the goal of sustainably developing economic activity and eradicating poverty in the Lake Victoria Basin. The scope between the Partner States was cooperation in the areas as they relate to the conservation and sustainable utilization of the resources of the Basin (Art. 3). The definition of water resources includes the groundwater as part of the living and nonliving resources therein (Art. 1). The Protocol follows the principles established by the 1997 UN Convention (Art. 4 of the Protocol) and also presents the need to establish an institutional framework and an organizational structure able to promote measures aimed at eradicating poverty and protecting the environment within the Lake Victoria Basin (Art. 33). This framework is formed by the Council of Ministers and establishes the basic functions for the Lake Victoria Basin Commission (ECOLEX 2003b).

11.5.7 Revised African Convention on Conservation of Nature and Natural Resources (2003)

This Convention is a regional treaty and was adopted in Maputo, Mozambique, on July 11, 2003. Currently, 42 countries have signed it; 16 ratified it and entered into force on April, 2016 (African Union 2003). The Convention lays down obligations to conserve and to make sustainable use of groundwater resources, including wetlands. Article 7 specifically focuses on water resources, with the basic goals of development, management, and conservation of transboundary groundwater in a cooperative manner. Another important aspect is to prevent cross-border harm from pollution and from excessive extraction of groundwater (Art. 7.2). The principle of collaboration (Art. 12) is manifested through measures having a cross-border effect and establishing joint commissions (African Union 2003).

11.6 Conclusion

Transboundary groundwater is a vital source, which should be protected around the world. International law has recently adopted legal instruments addressing the management of transboundary groundwater. The analysis of treaty practices in Africa reveals that the trend is to regulate transboundary groundwater although only few agreements include as the main scope the regulation and management of this resource. Those treaties specifically regulating as their main scope transboundary aquifers basically focus on exchange of information, monitoring, and scientific approach. On another perspective, most treaties regulating transboundary watercourses include groundwater as a source connected to the transboundary watercourse and therefore applying the same measures to both sources similar to the 1997 UN Watercourses Convention, although the integration of both sources can be difficult to manage in a context of water stress and water scarcity. This analysis also reveals that most legal instruments reflect the principles of international water law, although the implementation of them can be difficult, and thus, additional cooperation mechanisms such as commission at the local level would ensure the effective sustainability of groundwater resources in Africa.

References

AFB (2015) Water supply & sanitation in Africa: findings, lessons and good practices to improve delivery. African Development Bank. Available at http://idev.afdb.org/en/document/water-supply-sanitation-africa-findings-lessons-and-good-practices-improve-delivery-0

African Union (2003) African convention on conservation of nature and natural resources. Retrieved May 7:2008. http://www.fao.org/faolex/results/details/en/c/LEX-FAOC045449

Altchenko Y, Villholth KG (2013) Transboundary aquifer mapping and management in Africa: a harmonised approach. Hydrogeol J 21:1497–1517

Barberis JA (1986) International groundwater resources law, vol 40. Food & Agriculture Org, Rome

Burchi S, Mechlem K (2005) Groundwater in international law: compilation of treaties and other legal instruments, vol 86. Food & Agriculture Org, Rome, Italy

Camacho Suarez V, Saraiva Okello A, Wenninger J, Uhlenbrook S (2015) Understanding runoff processes in a semi-arid environment through isotope and hydrochemical hydrograph separations. Hydrol Earth Syst Sci 19:4183–4199

CGIAR (2017) Transboundary aquifer map for Africa. https://wle.cgiar.org/content/transboundary-aquifer-map-africa. Accessed 4 Aug 2017

Davis SN, DeWiest R (1966) Hydrogeology. Wiley, New York, p 463

Eckstein GE (2005) A hydrogeological perspective of the status of ground water resources under the UN watercourse convention. Colum J Envtl L 30:525
Eckstein GE, Eckstein Y (2003) A hydrogeological approach to transboundary ground water resources and international law. Am U Int'l L Rev 19:201
Eckstein Y, Eckstein G (2004) Looking at ground water resources from opposite sides of the glass: the differing perspectives of hydrogeologists and lawyers on ground water resources. Paper presented at the Transboundary Groundwater Flow Symposium at XXXIII IAH Congress, Zacatecas, Mexico, October 11–15
ECOLEX (2002) Tripartite Interim Agreement between the Republic of Mozambique and the Republic of South Africa and the Kingdom of Swaziland for Co-Operation on the Protection and Sustainable Utilisation of the Water Resources of the Incomati and Maputo Watercourses. Environmental Law database online. www.ecolex.org. Accessed 2017
ECOLEX (2003a) Convention on the Sustainable Management of the Lake Tanganyika between Burundi, Democratic Republic of Congo and the United Republic of Tanzania. Environmental Law database online. www.ecolex.org. Accessed 2017
ECOLEX (2003b) Protocol for Sustainable Development of Lake Victoria Basin between the Republic of Kenya, the Republic of Uganda the United Republic of Tanzania, Burundi and Rwanda. Environmental Law database online. www.ecolex.org. Accessed 2017
FAOLEX (2000) Agreements between Chad, Lybia, Egypt and Sudan, No. 1 Agreement Concerning the monitoring and exchange of information related to the groundwater of the Nubian Sand-stone Aquifer System. And agreement No. 2 For monitoring and sharing data for the sus-tainable development and proper management of the Nubian Sandstone Aquifer System. FAO legal database online. Accessed 2017
Freeze RA, Cherry JA (1979) Groundwater. 47 Library of Congress Cataloging in Publication Data
Guarani Aquifer Agreement (2010) San Juan, Republic of Argentina
Harmon J (1895) Attorney-General, to Richard Olney, Secretary of State 12 Dec. 1895. 21 Op. Att'y Gen. 274 (1895), reproduced in U.S. APPENDIX, p 204
IAEA (2011) Nubian Sandstone Aquifer System. International Atomic Energy Agency. http://www-naweb.iaea.org/napc/ih/documents/factsheetsPosters/Nubian%20-%20Transboundary%20Aquifers%20and%20Rivers%20Basins.pdf. Accessed 30 July 2017
IAEA, UNDP–GEF (2013) Regional strategic action programme for the Nubian aquifer system, signed by Arab Republic of Egypt, Libya, Republic of the Sudan and Republic of Chad. International Atomic Energy Agency, United Nations Development Programme, Vienna
IGRAC (2015) The 2015 map of transboundary aquifers of the world. UNESCO International Hydrological Programme (IHP). https://www.un-igrac.org/resource/transboundary-aquifers-world-map-2015
IGRAC (2017a) Africa groundwater portal. International Groundwater Resources Assessment Centre. https://www.un-igrac.org/special-project/africa-groundwater-portal
IGRAC (2017b) Groundwater, the hidden resource. International Groundwater Resources Assessment Centre. https://www.un-igrac.org/special-project/groundwater-hidden-resource
ILC (1994) International Law Commission of the United Nations on the Work of its Forty-sixth Session, Resolution on Confined transboundary Groundwater. 1994 2 Y.B. INT'L L. COMM'N, pt. 2, at 135, UN Doc. A/49/10
ILC (2008) Report of the International Law Commission on the Work of its Sixtieth Session. UN GAOR, 63d Sess., Supp. No. 10, at 19, UN Doc. A/63/10
Krishna R, Salman S (1999) International groundwater law and the World Bank policy for projects on transboundary groundwater. World Bank Technical Paper:163–190
McCaffrey S (1991) The law of international watercourses, 2nd edn. Oxford University Press Inc, New York
McCaffrey SC (2006) Understanding International Law. LexisNexis, Atlanta
McCaffrey SC (2009) The International Law Commission adopts draft articles on transboundary aquifers. Am J Int Law 103:272–293
McCaffrey S (2010) Sovereignty and cooperative management of shared water resources in a time of shrinking availability: the role of international law. In: Transboundary Aquifers, Challenges and New Direction, Pre-proceedings of ISARM 2010 Conference. UNESCO IHP, Paris

Mechlem K (2016) Groundwater governance: the role of legal frameworks at the local and national level—established practice and emerging trends. Water 8:347

Moore JE, Zaporožec A, Mercer JW (1995) Groundwater, a primer, vol 1. Amer Geological Institute, Alexandria

OSS (2016) 2012 – 2015 Activities reports & prospects. Submitted by the Executive Board to the 5th Session of the General Assembly Item 2.1 – AG I Item 10.2 – CA Rabat. Sahara and Sahel Observatory. http://www.oss-online.org/en

Ruhl JF, Kanivetsky R, Shmagin B (2002) Estimates of recharge to unconfined aquifers and leakage to confined aquifers in the seven-county metropolitan area of Minneapolis-St. Paul, Minnesota. US Geological Survey, Mounds View

SADC (2000) Revised protocol on shared watercourses in the Southern African Development Commu-nity. Southern African Development Community, Windhoek

Sandoval-Solis S, McKinney DC, Teasley RL, Patino-Gomez C (2011) Groundwater banking in the Rio Grande Basin. J Water Res Pl-Asce 137:62–71. https://doi.org/10.1061/(Asce)Wr.1943-5452.0000091

Scheumann W, Herrfahrdt-Pähle E (2008) Introduction. In: Conceptualizing cooperation on Africa's transboundary groundwater resources. German Development Institute (DIE), Bonn. https://www.die-gdi.de/uploads/media/Studie_32.pdf

Semedo MH (2012) Message to readers. Nature Faune 27:1–3

UN (1997) United Nations Convention on the Law of Non-navigational Uses of International Watercourses UN Doc. A/RES/51/229

UN Treaty Collection (1992) Chapter XXVII Environment, Status of the Convention on the Protection and Use of transboundary Watercourses and International Lakes. Helsinki, March 17 1992

UN Treaty Collection (2017) Chapter XXVII Environment, Status of the Convention on the Law of the Non-Navigational Uses of International Watercourses. New York, May 21 1997

UNECE (1992) United Nation Convention on the Protection and Use of transboundary Watercourses and International Lakes vol 1936 U.N.T.S. 269. United Nations, Helsinki, March 17, 1992

UNECOSOC (1999) Econ. & Soc. Council, Economic Commission for Europe, Meeting of the Parties to the Convention on the Protection and Use of transboundary Watercourses and International Lakes. Amendment to Articles 25 and 26 of the Convention vol ECE/MP.WAT/14

USGS (2016) What is a watershed? United States Geological Survey. https://water.usgs.gov/edu/watershed.html. Accessed 6 July 2017

Index

E. O. Vieira et al. (eds.), *Integrated Water Resource Management*,
https://doi.org/10.1007/978-3-030-16565-9